Palgrave Global Media Policy and Business

Series Editors
Petros Iosifidis, Department of Sociology, City University, London, UK
Jeanette Steemers, Culture, Media & Creative Industries, King's College
London, London, UK
Gerald Sussman, Urban Studies & Planning, Portland State University,
Portland, OR, USA
Terry Flew, Creative Industries Faculty, Queensland University of
Technology, Brisbane, QLD, Australia

The *Palgrave Global Media Policy and Business Series* has published over 22 books since its launch in 2012. Concentrating on the social, cultural, political, political-economic, institutional, and technological changes arising from the globalization and digitization of media and communications industries, the series considers the impact of these changes on business practice, regulation and policy, and social outcomes. The policy side encompasses the challenge of conceiving policy-making as a reiterative process that recurrently addresses such key challenges as inclusiveness, participation, industrial-labor relations, universal access, digital discrimination, and the growing implications of AI in an increasingly global-ized world, as well as local challenges to global media business and culture. The business side encompasses a political economy approach that looks at the power of transnational corporations in specific contexts - and the controversies associated with these global conglomerates. The business side considers as well the emergence of small and medium media enterprises, and the role played by nation-states in promoting particular firms and industries.

Based on a multi-disciplinary approach, the series tackles four key questions:

- To what extent do new developments in platforms, and approaches to personal data require radical change in regulatory philosophy and objectives towards the media?
- To what extent do technologies, datafication and transforming media consumption require fundamental changes in business practices and models?
- To what extent do privatisation, datification, globalisation, and commercial-isation alter the creative freedom, cultural and political diversity, values and public accountability of media enterprises?
- To what extent does the structure of global communications contribute to (in)equality within the Global South?

Series Editors
Professor Petros Iosifidis, City, University of London, UK, p.iosifidis@city.ac.uk
Professor Jeanette Steemers, King's College London, UK, jeanette.steemers@kcl.ac.uk
Professor Gerald Sussman, Portland State University, USA, sussmag@pdx.edu
Professor Terry Flew, The University of Sydney, Australia, terry.flew@sydney.edu.au
Book proposals should be submitted to **p.iosifidis@city.ac.uk**

More information about this series at
https://link.springer.com/bookseries/14699

Terry Flew · Fiona R. Martin
Editors

Digital Platform Regulation

Global Perspectives on Internet Governance

Editors
Terry Flew
Media and Communication
The University of Sydney
Sydney, NSW, Australia

Fiona R. Martin
Media and Communication
The University of Sydney
Sydney, NSW, Australia

ISSN 2634-6192 ISSN 2634-6206 (electronic)
Palgrave Global Media Policy and Business
ISBN 978-3-030-95219-8 ISBN 978-3-030-95220-4 (eBook)
https://doi.org/10.1007/978-3-030-95220-4

Cover illustration: © gremlin|Getty Images

This Palgrave Macmillan imprint is published by the registered company Springer Nature Switzerland AG
The registered company address is: Gewerbestrasse 11, 6330 Cham, Switzerland

Acknowledgments

This book is an outcome of the Australian Research Council funded Discovery Project *Platform Governance: Rethinking Internet Regulation as Media Policy* (DP190100222). We would like to thank our research and publishing assistants Ayesha Jehangir and Katharine Kirkwood for diligent tracking of contributions, our institutional colleagues in the Department of Media & Communications and further afield at the Australian Competition and Consumer Commission (ACCC) and Australian Consumer Communications Action Network (ACCAN), and Philip Schlesinger, Phil Napoli, Nick Couldry, Catharine Lumby, and Pawel Popiel for discussions and debate.

Praise for *Digital Platform Regulation*

"When it comes to the governance of digital platforms, the question of who is regulating whom is now serious enough that scholars have begun to refer to governments as platforms, too. This volume provides a valuable sampling of how this problem looks from the side of government, with cases from around the world ranging from particular tensions raised by specific industries and practices such as those of journalism to macro-level challenges for the nature of policy-making itself."

—Sandra Braman, *author of* Change of State: Information, Policy, and Power; *Texas A&M University*

"Internet regulation is now a space where everything is in question. This illuminating collection challenges not only the rule-setting powers of the global digital platforms, but also many of the key assumptions which media and internet studies scholars bring to the field. Timely, wide-ranging, and brimming with new ideas, this book will be an essential resource for students, scholars, and policy practitioners."

—Julian Thomas, *RMIT University, Australia*

CONTENTS

NOTES ON CONTRIBUTORS

Sven Brodmerkel is an Assistant Professor for Advertising and Integrated Marketing Communications at Bond University/Australia. Before his academic appointment, he worked as a communication strategist in the German advertising industry, consulting for a diverse range of high-profile companies, including T-Mobile International, Bayer AG, Novartis International, Nestlé Purina, and Nokia. His research focuses on the practices, politics, and ethics of new media technologies in the context of advertising and branding. He currently also investigates the changing nature of professional creativity and workplace cultures in the advertising industry in relation to various age, gender, and race inequalities.

Nicholas Carah is an Associate Professor in Media and Communication at the School of Communication and Arts at the University of Queensland. His research examines the advertising model and promotional culture of digital platforms. He is an Associate Investigator in the Centre of Excellence in Automated Decision-Making and Society and a Director of the Foundation for Alcohol Research and Education. His current projects investigate harmful industries marketing on digital platforms, including the use of digital platforms by alcohol and nightlife marketers, and the significance of machine vision in the algorithmic culture of social media platforms like Instagram.

Stuart Cunningham is Distinguished Emeritus Professor, Queensland University of Technology. He is well known for his research on topics such as emerging digital industries, the creative industries, and media and cultural policy. His recent books are *Media Economics* (with Terry Flew and Adam Swift, 2015) *Social Media Entertainment: The New Intersection of Hollywood and Silicon Valley* (with David Craig, 2019), *A Research Agenda for Creative Industries* (edited with Terry Flew, 2019), *Wanghong as Social Media Entertainment in China* (with David Craig and Jian Lin, 2021) and *Creator Culture: Studying the Global Social Media Entertainment Industry* (edited with David Craig, 2021).

Edward Hurcombe is a Research Associate at QUT. He researches how news and journalistic practice are transforming in relation to the technologies, economies, and user cultures of social media platforms. He is the author of *Social News: How Born-Digital Outlets Transformed Journalism* (Palgrave, forthcoming).

Oliver Eklund is a Ph.D. candidate and Research Assistant at Queensland University of Technology's Digital Media Research Centre (DMRC) where he studies Internet-distributed television and the evolution of cultural policy.

Terry Flew is Professor of Digital Communication and Culture at the University of Sydney. His books include *The Creative Industries, Culture and Policy* (SAGE, 2012), *Global Creative Industries* (Polity, 2013), *Media Economics* (Palgrave, 2015), *Understanding Global Media* (Palgrave, 2018), and *Regulating Platforms* (Polity, 2021).

He was President of the International Communications Association (ICA) from 2019 to 2020 and was elected an ICA Fellow in 2019. He is a Fellow of the Australian Academy of the Humanities (FAHA). He has advised companies including Facebook, Cisco Systems and the Special Broadcasting Service, and government agencies in Australia and internationally, including the Australian Communication and Media Authority and the Singapore Broadcasting Authority. He has held Visiting Professor roles at City University, London, and George Washington University, and is currently a Distinguished Professor with the Communications University of China and an Honorary Professor at the University of Nottingham Ningbo China.

Rosalie Gillett is a Postdoctoral Research Fellow in the QUT Node of the ARC Centre of Excellence for Automated Decision-Making and Society and Research Fellow in the Digital Media Research Centre at

QUT. Rosalie's research lies at the intersection between online gender-based violence and platform governance. She investigates women's experiences of online harm and safety, digital platforms' self-regulation practices, and what data platforms need to build automated tools that enable their users to feel safe. Rosalie was nationally recognized as a leading early-career thinker and communicator being awarded an ABC Top Five humanities media residency in 2021.

Lelia Green is Professor of Communications in the School of Arts and Humanities, Edith Cowan University, Perth. She has been researching the social implications of technological change for over 30 years, including her work as a Chief Investigator for two Australian Research Council Centres of Excellence: the ARC Centre of Excellence for Creative Industries and Innovation (2005–2013) and the ARC Centre of Excellence for the Digital Child (from 2021). Among other publications, she is author of *The Internet: An Introduction to New Media* (Berg, 2010) and *Technoculture: From Alphabe to Cybersex* (Allen & Unwin, 2001) and lead editor of *Framing Technology: Society, Choice and Change* (Allen & Unwin, 1994).

Amélie P. Heldt is a Researcher and Ph.D. candidate at the Leibniz Institute for Media Research|Hans-Bredow-Institut, Hamburg, and associated with the Humboldt Institute for Internet and Society, Berlin. In her research and her Ph.D. project, she focuses on the effects of freedom of expression in the digital sphere. She works on platform regulation, social media governance, the effects of new technologies on opinion formation and public discourse as well as the exercise of fundamental rights in the context of algorithmic decisions and autonomous systems.

Viet Tho Le is a Professional Journalist and Media Researcher. He works as a Research Assistant in the School of Arts and Humanities, Edith Cowan University, Perth, where he obtained his Ph.D. in a Media Studies, Communication and Journalism-related area in 2020. He has worked as a Senior Digital Journalist with the BBC, a Producer at SBS, and was Managing Editor at Binh Dinh newspaper in Vietnam. His research interests focus on political communication, new media, cultural studies, social media and young people, and digital citizenship.

Ramon Lobato is Associate Professor (Australian Research Council Future Fellow) in the School of Media and Communication, RMIT University. A screen industries researcher with a special interest in digital distribution, he is the author/editor of four books including *Netflix*

Nations (NYU Press, 2019), *Geoblocking and Global Video Culture* (INC, 2016), and *Shadow Economies of Cinema* (BFI, 2012). He co-leads the Global Internet TV Research Consortium and serves on the editorial board of *International Journal of Digital Media and Policy*, *Media Industries Journal*, and *International Journal of Cultural Studies.*

Fiona R. Martin is Associate Professor from the Department of Media and Communication at the University of Sydney and researches digital journalism, dialogic technologies, and platform regulation. Her books include *Sharing News Online*, with Tim Dwyer (Palgrave Macmillan, 2019), and *Mediating the Conversation* (Routledge, 2022).

James Meese is a Senior Lecturer at RMIT University and an Associate Investigator with the ARC Centre of Excellence for Automated Decision Making and Society. His research interests include media and telecommunications policy, journalism, and intellectual property. His two books are *Authors, Users, Pirates: Subjectivity and Copyright Law* (MIT Press) and *Death and Digital Media* (Routledge, co-authored).

Philip M. Napoli is the James R. Shepley Professor of Public Policy in the Sanford School of Public Policy at Duke University, where he also directs the DeWitt Wallace Center for Media & Democracy. He is the author/editor of seven books and over 60 journal articles. He is also a frequent contributor to publications such as *Wired*, the *Hill*, and the *Columbia Journalism Review*. He has conducted research on local news ecosystems and local news audiences for organizations such as the Democracy Fund, the Knight Foundation, the Cleveland Foundation, the Federal Communications Commission, Internews, Facebook, the New America Foundation, the Benton Foundation, and the Geraldine R. Dodge Foundation. He has engaged in research collaborations with nonprofit organizations such as Free Press, the Center for Creative Voices in Media, the Norman Lear Center, and the Minority Media & Telecommunications Council. His work has been featured in media outlets such as the *New York Times*, National Public Radio, the *Wall Street Journal*, and *Politico*.

Victor Pickard is the C. Edwin Baker Professor of Media Policy and Political Economy at the University of Pennsylvania's Annenberg School for Communication, where he co-directs the Media, Inequality and

Change (MIC) Center. He has published six books, including the award-winning *America's Battle for Media Democracy* and *Democracy Without Journalism? Confronting the Misinformation Society.*

Pawel Popiel is a Postdoctoral Fellow at the Media, Inequality and Change (MIC) Center at the University of Pennsylvania and Rutgers University. His research focuses on communication policy, particularly how politics shapes the oversight and regulation of digital media and emergent technologies and with what consequences. His work has been published in journals like *Critical Studies in Media Communication, Journal of Digital Media & Policy,* and the *Journal of Broadcasting & Electronic Media.* He obtained his Ph.D. from the University of Pennsylvania's Annenberg School for Communication.

Asa Royal is an Associate-in-Research at Duke University's DeWitt Wallace Center for Media & Democracy. There, he conducts research into the health of local news ecosystems, exploring the population dynamics of news outlets and investigating the proliferation of hyperlocal sources of misinformation. Drawing on his previous career as a software engineer, he also contributes to several software engineering efforts at the DeWitt Wallace Center, most recently helping to build a self-updating archive of manipulated media and associated fact checks. In his free time, he works with a Georgia-based nonprofit to construct and maintain a registry of all local elections in the state.

Alexa Scarlata is a Research Assistant at RMIT University and a Ph.D. candidate in the School of Culture and Communication at the University of Melbourne. Her research considers the achievements and constraints evident in the recent history of online television in Australia, particularly as this pertains to local drama production and policy. She is the Book Reviews Editor of the *Journal of Digital Media and Policy* and has published in *Critical Studies in Television, International Journal of Digital Television, Communications: The European Journal of Communication Research, Media International Australia* and *Continuum.*

Chunmeizi Su is a Postdoctoral Research Associate at the University of Sydney, Department of Media and Communications. Her research interests are mainly focused on platform studies, Chinese telecommunication companies (Baidu, Alibaba, and Tencent), screen industry studies, and cultural soft power. Her publications include book chapters 'The Will to Power: BAT in and Beyond China', published in *Willing Collaborators*

Foreign Partners in Chinese Media (Rowan and Littlefield, 2018), as well as journal articles on *Television and New Media* (2022), *Multicultural Discourses* (2019), and *Global media and Communication* (2021). Her most recent book *Douyin, TikTok, and China's Online Screen Industry: The Rise of Short Video Platforms* will be published by Routledge in 2023.

Nicolas Suzor researches the regulation of networked society. He is a Professor at the Law School at Queensland University of Technology and a Chief Investigator of QUT's Digital Media Research Centre. He is also a Chief Investigator of the ARC Centre of Excellence for Automated Decision-Making and Society. He is a member of the Oversight Board, an independent organization that hears appeals and makes binding decisions about what content Facebook and Instagram should allow or remove, based on international human rights norms. His research examines the governance of the Internet and social networks, the regulation of automated systems, digital copyright, and knowledge commons. He is the author of *Lawless: The Secret Rules that Govern Our Digital Lives* (Cambridge, 2019).

Dwayne Winseck is a Professor at the School of Journalism and Communication, with a cross-appointment to the Institute of Political Economy, Carleton University, Ottawa, Canada. His research interests include the political economy of telecommunications, the Internet and media as well as communications and media history, theory, policy, and regulation. He is also the Director of the Social Sciences and Humanities Research Council-supported Canadian Media Concentration Research Project and the Global Media and Internet Concentration Project.

List of Figures

CHAPTER 1

Introduction

Terry Flew and Fiona R. Martin

In recent years, questions of internet governance and digital platform regulation have moved from being a specialised niche field within internet and digital media studies to being at the forefront of scholarly, policy and community debate. The triggers for this have been many and varied. The 2013 Edward Snowden revelations about the extent of U.S. National Security Agency (NSA) monitoring of not only U.S. citizens, but platform users and political figures around the world. The close connections revealed between the NSA and many of the world's leading digital technology companies, made it clear that there was no structural separation in practice between the global internet, private capital and the surveillance agencies of democratic nation-states. This influenced Europe's eventual adoption of its General Data Protection Regulation (GDPR) (Rossi, 2018). From this time, national governments gave increasing attention to setting their own rules about data sovereignty and privacy rights, as Brazil

T. Flew (✉) · F. R. Martin
Media and Communication, The University of Sydney, Sydney, NSW, Australia
e-mail: terry.flew@sydney.edu.au

F. R. Martin
e-mail: fiona.martin@sydney.edu.au

© The Author(s) 2022
T. Flew and F. R. Martin (eds.), *Digital Platform Regulation*,
Palgrave Global Media Policy and Business,
https://doi.org/10.1007/978-3-030-95220-4_1

1

did in passing its *Marco Civil da Internet* (Brazilian Civil Rights Framework for the Internet) in 2014, and as Canada did from 2015 with data localisation laws. This period also saw the birth of the global Indigenous data sovereignty movement (Kukutai & Taylor, 2016) and the European Union taking an increasingly activist role in internet regulation within its jurisdiction, first with the GDPR implementation in 2018, and then the Digital Services Act and Digital Markets Act, tabled in 2020 and designed to set stricter rules and accountability frameworks for what the European Commission termed Very Large Online Platforms (VLOPs).

Evidence of platform companies' significant market power and their negative impacts on competition have spawned long-running anti-trust cases, such as the EU's much debated action against Google's shopping search discrimination (Eben, 2018; Schechner, 2021) and led to calls for tech company breakups (Gilbert, 2021). There are certainly industry commentators who have questioned attempts to define platforms for competition purposes (O'Connor & Schruers, 2016). However, there is now academic consensus that they constitute dominant software systems with a global reach that support multi-sided markets, linking geographically dispersed parties in trade, communication, social and cultural activity, and that there is a great need to better understand these organisational forms and their impacts in order to ensure they operate in, and within, the public interest (Gawer, 2021; Gorwa, 2019; Flew et al., 2019).

Notably, the tangible harms arising from social media misinformation, hate speech, terrorism, abuse and harassment have spurred both the introduction of punitive national laws, such as Germany's Network Enforcement Act (2017) and Australia's Sharing of Violent Abhorrent Material Act (2019), and major public inquiries, such as the U.S. Select Committee on Intelligence hearings about Russian influence on the 2016 election and the UK's 2017 Online Safety and 2021 Online Harms inquiries. While the growth of national initiatives to address social media harms has fuelled concerns about the emergence of a legal 'splinternet' and increased regulatory burden on platform operations, it is now apparent that governments see the future of safe, accountable, equitable internet communications and trade as reliant on new controls on platform power and influence. The November 2021 'Facebook papers' revelations that Facebook, Instagram and WhatsApp operations prioritised profit before public safety, amplifying instead of removing harmful content, often against employee advice, has intensified calls for greater regulatory oversight (Satariano, 2021).

Speaking at the Opening Plenary of the 2018 Internet Governance Forum, French President Emmanuel Macron flagged the need for a 'third way' in internet governance, between the perceived libertarianism of Silicon Valley and the authoritarian statism of the Chinese internet, arguing that platform regulation to restore accountability and trust was a pre-condition for maintaining the values of freedom and democracy associated with the early vision of the open internet (Macron, 2018). One approach to this can be seen in the European Commission's 2016 voluntary Code of Conduct on Countering Illegal Hate Speech Online, now in its fifth year (European Commission, 2021), which was introduced to stem a tide of abuse against immigrants during the 2015–2016 refugee mass migration to Europe. This content moderation monitoring governance exercise has seen all the major platforms cooperating with NGOs across Europe to act on their reports of hate speech, with the EC evaluating how well platforms are meeting their Code commitments.

The inspiration for this collection of essays came from such apparent paradigm shifts in understandings of the internet and its socio-economic and political role around the world. As the editors of this book, we are the beneficiaries of a research grant awarded by the Australian Research Council (ARC) through its Discovery Program, on Platform Governance: Rethinking Internet Regulation as Media Policy (DP190100222), along with Tim Dwyer and Chunmeizi Su (University of Sydney), Nicolas Suzor (Queensland University of Technology), Josef Trappel (University of Salzburg), and Philip Napoli (Duke University). We set out to explore the shifting balance between media policy and platform self-governance in the way digital platforms managed their commitments to both free speech and public wellbeing, and the issues arising from nation-state governance of platform content, including the prospects for developing international laws, norms and standards through multi-stakeholder approaches. We also hoped to develop an interdisciplinary understanding of how national media laws, systems and industry cultures continue to shape the practices and conduct of global platform companies operating in multiple jurisdictions.

During our research we were struck by the extent to which politicians, governments and regulators around the world had become increasingly activist towards the largest digital companies in particular, as part of what was described as the 'techlash' (*The Economist*, 2018) and the 'neo-Brandeisian' movement to revise antitrust laws to take on 'Big Tech' (sometimes also called 'hipster antitrust') (Khan, 2018; Rogoff, 2018;

Wu, 2018). It was also apparent that industry self-regulation, or 'regulation by public apology' (Hall, 2020; Tufecki, 2018), was seriously inadequate in the face of public shocks such as the Cambridge Analytica scandal that *The Observer and The Guardian* broke in 2018, and the livestreaming of the Christchurch Mosque shootings in 2019. Even Facebook (now Meta) CEO Mark Zuckerberg came to concede the need for regulation of businesses such as his own, acknowledging to the U.S. Congress in 2018 that 'the real question, as the Internet becomes more important in people's lives, is what is the right regulation, not whether there should be or not' (Zuckerberg & Senate Commerce, Science and Transportation Committee, 2018).

Yet this apparent regulatory embrace belies the effort platform companies have put into fighting attempts to regulate them. As far back as 2012 they successfully encouraged their global users to protest against the proposed U.S. Stop Online Piracy Act (SOPA) and the Protect IP Act (PIPA) (Benkler et al., 2015). They have marshalled support from free speech NGOs like the Electronic Frontier Foundation and Freedom House against proposals to force them to take more responsibility for what their users post and share. Both Google and Facebook have withdrawn services in response to new national legislation. Google withdrew its news services in Spain and Germany, following attempts by news publishers to negotiate payment for its display of their headlines and excerpts. Facebook withdrew services in Australia following the government's 2021 introduction of a News Media Bargaining Code (NMBC), banning Australians from accessing news via its platform, and the world from accessing Australian news accounts. Further, the structural complexity of platform eco-systems, and their interdependence with economic, political and social systems, has made traditional approaches to media and information regulation relatively ineffective, if not obsolete, and new governance strategies essential (Van Dijck, 2021).

There is thus a 'new regulatory field' (Schlesinger, 2020, p. 1558) which has emerged around digital platform companies' colonisation of the internet, our multi-faceted adoption of social media tools, and platforms' relentless 'datafication' of personal information (Meijas & Couldry, 2019). Dwayne Winseck has observed 'a dizzying number of public policy inquiries into the digital platforms' (Winseck, this volume), seeking to understand the scope of their influence, and their potential for harm. New laws, policies and regulations are being proposed by nation-states, in the liberal democracies as much as in less democratic states, that overlay an already complex web of standards, protocols, rules, regulations and governance structures which have been associated with the global internet since the 1990s (Mueller, 2017; Musiani et al., 2016). Even the United States, long the key advocate of the open, unregulated internet, has experienced an apparent paradigm shift, with the Biden administration giving key policy roles to critics of digital platform power such as Lina Khan and Tim Wu (Flew & Gillett, 2021), while erstwhile 'free market' advocates such as the Stigler Center at the University of Chicago call for stronger anti-monopoly laws in order to revive innovation in the digital economy (Stigler Center for the Study of the Economy and the State, 2019). Indeed the lines between democratic and authoritarian states are blurring in this space, with countries such as China adopting antitrust laws inspired by U.S. policy debates, in order to rein in the perceived market power of their own dominant platforms (Kasperkevic, 2021).

In this book, we focus on digital *communication* platforms. This is a difficult, yet necessary, distinction to make in discussion of platform regulation, given the platformisation of the internet (Helmond, 2015; Flew, 2019) has occurred in the wider context of the platformisation of business and trade more generally (McAfee & Brynjolfsson, 2017; Parker et al., 2016). There is a plethora of platform companies that are not in communications or media-related businesses, such as Uber, AirBnB, eBay and Upwork. However, big tech companies such as Google, Facebook and Microsoft are clearly in the businesses of communication, especially advertising, and interact with media companies in a sustained way, although this orientation may be less apparent for companies such as Apple and Amazon. Also, the technological lines between platforms and mainstream media have increasingly dissolved. Netflix has revolutionised television through the platformisation of content delivery, shaped by the analysis of connections between user preferences and behaviour, and the use of data and algorithmic selection based on behavioural targeting, which drives content commissioning decisions, but it does so in ways that are recognisably those of a media company (Lotz, 2021). In this instance all other media companies are increasingly looking like Netflix, with their on-demand and streaming video platforms (e.g. Disney+, BBC iPlayer) and data-driven decision-making processes.

While platform companies have long maintained that they are content hosts not publishers, and thus are not media companies in the traditional editorial sense, these lines have also been crossed. The Australian Competition and Consumer Commission (ACCC), in its 2019 *Digital Platforms*

Inquiry Final Report, observed that companies such as Google and Facebook increasingly perform 'media-like functions' of commissioning, editing, curating and distributing media content, thus giving them a key role in 'shaping the online news choices of Australian consumers' (Australian Competition and Consumer Commission, 2019, p. 173).

In this book, we are interested in the forms of regulation and governance that might be applied to those digital platforms which offer communications and media as a service. This includes both the broad-reach, multifaceted platforms such as Google and Facebook, but also the more narrowly marketed, sector-specific platforms such as Netflix. It includes companies such as Apple, Amazon and Microsoft insofar as they provide communication services and media publishing, such as iCloud, Twitch, LinkedIn and Yammer, or provide media content: think of Apple+ TV and the App Store, Amazon Video and Audible, or Microsoft's Xbox. The argument that such platforms *are* media and communications companies (Napoli & Caplan, 2017) is not without its critics: there is debate about this question in our collection (see Pickard and Winseck's chapters), as well as between authors such as Philip Napoli (2019) and Dwayne Winseck (2020).

The Clinton Administration's Section 230 of the *Communications Decency Act*, passed in 1996 and one of the very few articles from that legislation to survive a Supreme Court challenge, is commonly held to be the cornerstone of the platform/publisher distinction, with the idea that 'internet intermediaries' (as they were then known) may act to block, remove or downgrade content on their sites without acquiring the legal status of publishers, and cannot be held legally accountable for the content posted by their users (Gillespie, 2017). The argument does, however, go further back in the internet imaginary, with Ithiel de Sola Pool's foundational 1983 text, *Technologies of Freedom*, first crystallising the argument that new forms of electronic communication technologies required a 'policy of freedom' (de Sola Pool, 1983) that clearly demarcated them from regulation-bound print and broadcasting industries. He envisaged an internet that was inherently, in form and operation, resistant to legal constraint:

> Electronic media...allow for more knowledge, easier access and freer speech than were ever enjoyed before...one might anticipate these technologies of freedom will overwhelm all attempts to control them. (de Sola Pool, 1983, p. 251)

Yet early ideas of the internet as a frontier territory, unconstrained and uncontrollable by the rule of law, were never accurate given the numerous international governing bodies and intertwined governance processes that have been necessary to build and maintain the network of networks—and despite John Perry Barlow's (1996) proclamation to the contrary—never free of tyranny. Waves of regulatory concern about copyright, classification, child pornography, net neutrality and terrorism have generated continual debates about how we might best preserve the liberalising and innovative power of the internet, while acknowledging states' territorial sovereignty *and* enabling effective international regulatory efforts (Savin, 2017).

One of the conceptual challenges we have faced in defining activity in this field is whether we are talking about digital platform *regulation* or digital platform *governance*. The concept of regulation typically refers to actions by governments and public agencies on private actors that are enabled by binding laws and which have negative sanctions for non-compliance. Koop and Lodge define regulation as 'intentional intervention in the activities of a target population, where the intervention is typically direct – involving binding standard-setting, monitoring, and sanctioning – and exercised by public-sector actors on the economic activities of private-sector actors' (Koop & Lodge, 2017, p. 105). By contrast, governance – derived from the Latin verb *gubernare*, meaning 'to steer the ship' – is taken to be associated with a more decentred conception of where power and control lies, encompassing of both the agencies and activities which shape the conduct of actors such as private companies, including those companies' own attempts at self-rule. Mark Bevir has defined governance in these terms:

> Governance draws attention to the complex processes and interactions that constitute patterns of rule. It replaces a focus on the formal institutions of states and government with recognition of the diverse activities that often blur the boundaries of states and society. Governance … highlights phenomena that are hybrid and multijurisdictional with plural stakeholders who come together in networks. (Bevir, 2011, p. 2)

There is a certain natural affinity between the internet and digital platforms on the one hand, and governance practices based on rough consensus rather than formal rules on the other. At a conceptual level, the proposition that decision-making power flows through multiple decentralised networks, nodes and machines sits squarely with understandings of internet culture as being informed by actor-network theories (Latour, 2007), and its reliance on forms of collective coordination (Puppis, 2010) with notions of the internet driving a shift towards network organisations (Thompson, 2003), network economies (Benkler, 2006, 2011), and network societies (Castells, 1996, 2009, 2010, 2012). The internet's international institutions have never been understood as top-down entities able to impose rules on, and enforce sanctions against, nation-states. Rather, agencies such as ICANN and the Internet Governance Forum are seen as exemplifying principles of multistakeholder cooperation. The institutions involved in global internet governance are framed around tripartite institutional representation, bringing representatives of civil society organizations (NGOs, academics, etc.) and industry bodies to the table, either alongside governments or as an alternative to them (Bray & Cerf, 2020; DeNardis, 2014; Mueller, 2010). Tripartism and multistakeholder approaches have often been preferred frameworks for addressing issues with digital platform companies, such as guiding principles for content regulation, as they avoid the perceived risks of censorship associated with direct state involvement in making decisions in such domains.

At a more general level, governance relations are at the core of platform businesses. As they operate by definition in multi-sided markets, and since the guiding principle of their business model is to enable 'core interactions between platform participants, including consumers, producers, and third-party actors' (Constantinides et al., 2018, p. 381), these companies have to establish *ad hoc* governance arrangements in order to keep all participants and stakeholders engaged and satisfied with their performance and value-adding capacities. As Flew has observed elsewhere 'a platform without governance is not possible; governance is as central to platforms as are data, algorithms, and interfaces' (Flew, 2021, p. 135).

The breadth of the governance concept is, however, both its strength and weakness. It undoubtedly captures forms of practice which aim to shape the conduct of others without direct regulation. One thinks, for instance, of the many behavioural 'nudges' that are now central to contemporary public policy, where preferred outcomes are achieved by reshaping the 'choice architecture' of individuals rather than telling them what they must and/or cannot do (Halpern, 2015; Thaler, 2015). At the same time, governance-based approaches to reshaping the conduct of digital platforms invariably require corporate self-regulation, and raise the question of whether this internal oversight is sufficient to address issues of public concern, or whether it is time for governments to develop stronger rules that have meaningful sanctions for non-compliance.

As with debate about whether communications platforms are media companies, there is a lively debate in this collection about the pros and cons of platform self-regulation. Victor Pickard (this volume) argues that reliance upon corporate self-regulation and social responsibility is always going to be insufficient in the face of business models which promote monopolistic and ethically dubious practices, and that more radical structural reforms—such as the break-up of the big platforms—are required. Closely interrogating the concept of corporate social responsibility (CSR), Lelia Green and Viet Tho Le argue that it can only play a meaningful role if accompanied by state regulation. By contrast, Nicolas Suzor and Rosalie Gillett (this volume) argue that the content moderation decisions of digital platforms will always require a degree of discretion, and that platform self-regulation is always going to be a part of the regulatory mix, even if there are also moves towards more direct government regulation. This is because 'content moderation and curation is the commodity that platforms offer to their users' (Suzor and Gillett, p. 274), and the different approaches that they adopt in shaping these governance arrangements is inevitably a part of their business model and the contract they offer to their users and multiple stakeholders.

Platform companies' increasingly tight grip on digital advertising spend, and the resulting dire consequences for both democratic communications and a news media industry already wounded by plummeting circulation and increased competition, have motivated intense regulatory debate. As the UK's Cairncross Review argued, the platformisation of news has sponsored market failures with declines in local reporting, political coverage and expensive investigative journalism. Cairncross too found the "unbundled" experience of platform news encounters was having negative impacts on the "visibility of public-interest news and for trust in news" (2019, p. 6). With this in mind, we open our collection with reflections on the types of regulation that might counter the incursions of search and social media platforms on the advertising revenue that once supported journalism.

In the first of our contributions to this collection, North American media studies researcher Victor Pickard explores systemic approaches to supporting public interest journalism in the platform era, ranging from platform company levies to publicly funded media alternatives. He suggests that platforms' profit first focus, their adherence to an apolitical "marketplace of ideas" conception of free speech, and their embedding in a North American discourse of negative freedoms (ie. against regulation) mean they are unlikely to self-address the structural inequities in voice and influence they entrench. Yet even as Pickard characterises platform companies as "vertically integrated monstrosities, wielding a degree of political power incompatible with a functioning democracy", he also rejects a resort to anti-monopolist, corporate breakup scenarios. Instead he favours solutions that not only curb platform power to determine news agendas, but also ameliorate the commercial drift of digital publishers to sensationalist, click-driven, socially irresponsible reporting. Amongst those he canvases are what in the neo-liberal moment might be seen as 'radical' alternatives: action from unionised platform workers, legislation that regards platforms as public utilities, and the potential creation of public social media.

U.S. media regulation scholars Philip Napoli and Asa Royal then take up the issue of the fraught relationship between platforms and news publishers from a different perspective: that of the press' legal and political battles to wrest compensation from platform companies for the snippets of news content they display and their users re-distribute. In this account, which reviews long running copyright cases in France and Germany, the EU's Directive on Copyright in the Digital Single Market has opened the door to at least one content licensing deal, but with terms that are opaque, and which do not acknowledge publishers' rights in their excerpts. In its coverage of Australian case, based on competition law, the chapter notes how government attempts to mandate platform-publisher negotiations over the value of news led to Facebook's infamous news ban, demonstrating both its market power and its disregard for civil society. Here too, as in France, we see that deals with Google and Facebook under the NMBC lack transparency and benefit larger companies or those that bargain collectively. While concluding that government intervention seems essential to secure the future of news journalism, Napoli and Royal's chapter also suggests the difficulty of approaching platform regulation from isolated, issue-based perspectives. In this respect, an integrated approach to media reform of the type proposed by the ACCC's *Digital Platforms Inquiry* can likely return better outcomes that individual legislative changes or dependence on platform self-regulation and industry support.

The need for a coherent program of reforms to meet the challenges of digitalisation and platformisation, is part of the narrative legal scholar Amélie P. Heldt presents in reviewing platform obligations under the EU's proposed Digital Services Act (DSA), and how these are monitored for compliance. Heldt notes that the driving force for the Act was member states individual moves to legislate against online harms, a patchwork of legislation that suggested the EU needed a more uniform approach to intermediary liability and user safety, and rules for removal of illegal content. Under the DSA, platforms are also obligated to provide feedback to the source of removed illegal content about the rationale for its erasure, and an internal complaints handling process for users more broadly, moves the platforms have resisted due to the administrative burden of compliance. However, this aspect of the DSA does not address a key finding of the EC's fifth hate speech monitoring trial, which found platforms also need to improve their feedback to users who notified them of illegal content, detailing actions taken (Reynders, 2020) a move which would encourage more effective flagging. Where the DSA does innovate, according to Heldt, is in the establishment of two new regulatory agents, national Digital Services Coordinators and a regional Board for Digital Services, which will work in tandem with the European Commission, and in mandating that platform companies abide by the EU Charter of Fundamental Rights in their dealings with their services' users and competitors.

The question of platform companies' 'social license to operate', and their responsibilities to the societies and communities they serve, has been brought sharply into relief by Facebook's use in the 2018 genocide of the Rohingya minority (Lee, 2019), and more recently social media's contribution to the 2021 storming of the U.S. Capitol (Schewe, 2021). In their chapter, communications scholars Lelia Green and Viet Tho Le use former President Donald Trump's deplatforming after the Washington D.C. Capitol riot on January 6, 2021 as a jumping off point to reflect on the types of social responsibility we might expect from platform companies, as well as the regulatory measures and civic action that might encourage them to better address social concerns and democratic principles. Certainly, we have seen increasing platform attention to responsibility in advertising since the establishment of GARM, the Global Alliance for Responsible Media, a World Federation of Advertisers move to explore the mitigation of "harmful content on digital media platforms and its monetization via advertising" (GARM, 2021) and the 2020 international #StopHateforProfit campaign mobilised Coke and Unilever to support its protest. However, the debate Green and Le engage about what constitutes good corporate citizenship in content publishing and moderation underscores the extent to which platforms are making, via AI filtering and/or rapid human assessment, even more significant editorial decisions once taken by licensed media companies and monitored by national agencies. As Van Dijck et al. (2021) argue, their move to deplatformisation, the wholesale preventative removal of dangerous individuals and their organisational networks, "exposes an accountability gap" between them, governments and public. It is precisely this type of power they argue, which controls access to the essential infrastructure of global communicative participation, that demands more transparent regulatory intervention at national and supranational levels.

In this respect digital media researchers Nicolas Carah and Sven Brodmerkel, in our collection, present a persuasive case that we also need to know far more about the forms, impacts and consequential harms of platformised advertising and the influence this imparts platform companies, given Google and Facebook's share of digital advertising spend globally accounts for 28.6 and 23.7 per cent respectively in 2021 (eMarketer, 2021). Using the case of online alcohol marketing, and its new participatory data fuelled platform model, this case study contributes significantly to our knowledge of how platform companies have transformed advertising and ad markets through data analytics and interface design. Their algorithmic brand cultures not only micro-target advertising to user preferences and behaviour, but also encourage vernacular creativity from influencers and users to boost campaign impact. While historically advertising regulation has been concerned with representation of drinking cultures, now they argue we should be more concerned about the opacity of advertising's reach and influence, the difficulty of understanding who has been targeted, with what, and with what consequences for public health and other socially beneficial outcomes.

Throughout this collection, the contributing authors provide a lively critique of platform companies' resistance to administrative transparency, and their reluctance to reveal exactly how they intervene in public debates, or what they do to mediate dangerous and risky content. In communications scholar Pawel Popiel's chapter he provides us with a new lens on transparency, by tracking how the major U.S. platform companies try to influence policy debates: the issues they tackle, the policy approaches they favour, and what their policy communications suggest about their attitudes to regulation and governance. His analysis confirms their interest in technological solutionism, and what he calls "frictionless regulation", the self-defined, rapidly evolving territory of platforms' community standards and issue-based (often seemingly ad-hoc) multi-stakeholder engagement. This focus, he argues simultaneously advances their business interests while avoiding structural interventions into their operations or entanglement in lengthy public deliberations as the ACCC's Rod Simms told European policy-makers recently: "what we've observed…is that Facebook and Google, they really just do things on "take it or leave it" terms. They dictate the terms of the arrangement" (Sims, 2021). So while platform companies may accede to national co-regulation in certain areas such as data privacy, Popiel warns that they will move fast to set the policy agenda, with the worrying possibility of state capture by private interests.

A micro-analysis of Facebook and Google's policy agency, by James Meese and Edward Hurcombe, then reveals how this dynamic played out during the formulation and introduction of the ACCC's News Media and Digital Platforms Mandatory Bargaining Code. Here, we see a regulatory action that sought to make big platforms pay for news, but which avoided designating either company as actionable under the new law because they both negotiated deals with publishers before that happened. Meese and Hurcombe undertake a close read of the policy process to challenge the common view that the Code benefited big media rather than journalism or media diversity (see Warren, 2021). Their decentred analysis of institutional alignments in industry and political agendas reveals how ongoing stakeholder negotiations led to new regulatory obligations on Google and Facebook, despite their apparent economic power. Their account highlights the need for situationally and historically nuanced accounts of policy development that consider path dependencies as factors in regulatory outcomes.

Chunmeizi Su then takes up this challenge, exploring how Australia might differently consider regulating the activities of North America's tech giants and their Chinese counterparts Baidu, Alibaba, and Tencent, in light of the latter groups' growing base of Chinese-Australian users. She notes that while both Facebook and WeChat have generated initiatives to combat misinformation, WeChat is less likely to trigger direct government responses in Australia as it is principally a platform for the Chinese diaspora, whereas Facebook is closer to being a 'mass' communication medium.

The chapter focus then turns to the fate of local cultural production markets in an era of platformisation, another critical concern for policy makers with the rise of global subscription video-on-demand (SVOD) streaming services like Netflix and Disney+. Stuart Cunningham and Oliver Eklund highlight the competitive and information asymmetries between the highly regulated, territorially-bound broadcast sector and the relatively unregulated, unbound digital video "curation, aggregation and sharing" sector which have enabled SVOD companies to act as market disruptors in the screen industries, drawing parallels between the regulatory challenges raised by the market dominance of search and social media platforms and those of streaming platforms. Using three case studies, Cunningham and Eklund trace how European, Canadian and Australian regulators have sought to implement digital media policy reform that meets competition, social, cultural and public interest information goals, and differently address the contentious proposition that platforms should contribute financially to local cultural production in return for market access.

Applying a closer lens to the "Netflix Effect", or the influential market impact of its algorithmic production model, Ramon Lobato and Alexa Scarlata then investigate 'discoverability', a key aspect of this model, and its implications for media and information policy. Discoverability, or the mechanisms that act to make content visible to streaming platform users, has become a hot button policy issue due to the potential for some sources and types of content formerly privileged in legacy policy (such as local, minority language and documentary content) to be marginalised on streaming services. In exploring the breadth of editorial and system design factors that govern how content recommendations are made, Lobato and Scarlata rehearse the distributive politics of visibility and then unpack their realisation in national media policies of Canada, the UK, Australia and the European Union. Importantly they question the transparency and contestability of decision-making which shapes the prominence of competing channels and public service media content in streaming delivery.

The preceding two chapters position communications platforms comfortably within the ambit of existing media policy and regulation, which more or less is the proposal that has underpinned our research over the past two years. In contrast, telecommunications researcher and political economist Dwayne Winseck argues that trying to shape platform behaviour along broadcasting principles is mere political expedience, and ignores tech companies' closer historical alignment with telecommunications, electronics and finance sectors. For Winseck a pre-history of digital information networks suggests four principles on which we should base any future regulatory moves on platform companies: structural separation of large corporations; line of business restrictions; the imposition of public interest obligations and the provision of public service media *and* communications alternatives.

Whatever the policy framework that we may wish to apply to the conundrums of platform regulation and governance, the question of what role self-regulation and corporate governance should play looms close in a political climate dominated in the West by 'light touch' regulatory approaches, neoliberal economics and populist governments. Media law scholars Nicolas Suzor and Rosalie Gillett argue that persuading platform companies to exercise better self-regulation is a critical part of any oversight framework. After consulting a variety of regulatory experts, they argue that self-regulation provides: faster, more flexible, informal means of enforcing content standards, and acting to remove harmful material although these may suffer from a legitimacy-deficit. They also canvas the problems that civil society actors have in influencing platform decision-making and note the need for more effective platform consultation of government and civil society. Critically they note the difficulty of external parties influencing longer term, significant policy directions.

Despite the clear need for platforms to improve their self-governance, at this moment the politics of self-regulation are somewhat on the nose, especially in the wake of Facebook and Instagram's struggles with COVID19 misinformation and especially since the release of the Facebook papers with their spectacular expose of Meta's internal policy discontents. It seems fitting then that our final contribution from Terry Flew turns an interdisciplinary lamp on the reasons why debates about tech policy have wandered for decade in the discourse of governance, and now are turning regulatory with some fervour. Building on research into electoral swings and the rise of populist governments, Flew argues that technology policy, once the province of cosmopolitan tech-savvy elites, is now yielding to more conservative forces – leaving information activists torn between options that might curb human rights harms, but may equally curtail free speech.

As the European Union, the Brexited UK, and Canada look to introducing new platform-oriented policy reforms, this collection provides invaluable insights into the lenses that can be applied to those deliberations. It canvases the variety of stakeholders that require consideration and the intricacies of their relationships, gaps in regulatory research and the complexity of the field as it emerges. Thanks to our geographical location, this work certainly foregrounds activity in Australia and its region, but we regard this as an important balance to global north perspectives, and a worthy focus on the shift to national regulatory activism that is informing approaches in Europe and elsewhere.

References

Australian Competition and Consumer Commission. (2019). *ACCC Digital Platforms Inquiry Final Report*. ACCC. https://www.accc.gov.au/focus-areas/inquiries/digital-platforms-inquiry.

Barlow, J. P. (1996). A Declaration of the Independence of Cyberspace. Electronic Frontier Foundation. https://www.eff.org/cyberspace-independence.

BBC. (2020, September 23) Advertisers Strike Social Media Deal Over Hate Speech. *BBC News*. https://www.bbc.com/news/technology-54266534.

Benkler, Y. (2006). *The Wealth of Networks: How Social Production Transforms Markets and Freedom*. Yale University Press.

Benkler, Y. (2011). A Free Irresponsible Press: Wikileaks and the Battle over the Soul of the Networked Fourth Estate. *Harvard Civil Rights Civil Liberties Law Review, 46*, 311–398.

Benkler, Y., Roberts, H., Faris, R., Solow-Niederman, A., & Etling, B. (2015). Social Mobilisation and the Networked Public Sphere: Mapping the SOPA-Pipa Debate. *Political Communication, 32*(4), 594–624.

Bevir, M. (2011). Governance as Theory, Practice and Dilemma. In M. Bevir (ed.), *The SAGE Handbook of Governance* (pp. 1–16). SAGE.

Bray, D., & Cerf, V. (2020). The Unfinished Work of the Internet. In W. H. Dutton & M. Graham (eds.), *Society and the Internet* (2nd ed., pp. 403–417). Oxford University Press.

Cairncross, F. (2019). The Cairncross Review: A Sustainable Future for Journalism. UK Government, London. https://www.gov.uk/government/publications/the-cairncross-review-a-sustainable-future-for-journalism.

Castells, M. (1996). *The Rise of the Network Society: The Information Age: Economy, Society and Culture*. Blackwell.

Castells, M. (2009). *Communication Power*. Oxford University Press.

Castells, M. (2010). The New Public Sphere: Global Civil Society, Communications Networks, and Global Governance. In D. K. Thussu (ed.), *International Communications: A Reader* (pp. 36–47). Routledge.

Castells, M. (2012). *Networks of Outrage and Hope: Social Movements in the Internet Age*. Polity Press.

Constantinides, P., Henfridsson, O., & Parker, G. (2018). Platforms and Infrastructures in the Digital Age. *Information Systems Research, 29*(2), 381–400.

DeNardis, L. (2014). *The Global War for Internet Governance*. New Haven: Yale University Press.

de Sola Pool, I. (1983). *Technologies of Freedom*. Cambridge, MA: Harvard University Press.

Eben, M. (2018). Fining Google: A Missed Opportunity for Legal Certainty? *European Competition Journal, 14*(1), 129–151. https://doi.org/10.1080/17441056.2018.1460973.

eMarketer. (2021) Net Digital Ad Revenue Share Worldwide, by Company, 2019–2023 (% of Total Digital ad Spending). *Insider Intelligence*. https://www.emarketer.com/content/duopoly-still-rules-global-digital-ad-market-alibaba-amazon-on-prowl.

European Commission. (2021). The EU Code of Conduct on Countering Illegal Hate Speech Online. https://ec.europa.eu/info/policies/justice-and-fundamental-rights/combatting-discrimination/racism-and-xenophobia/eu-code-conduct-countering-illegal-hate-speech-online_en

Flew, T. (2019). The Platformized Internet: Issues for Internet Law and Policy. *Journal of Internet Law, 22*(11), 4–16.

Flew, T. (2021). *Regulating Platforms*. Polity Press.

Flew, T., & Gillett, R. (2021). Platform Policy: Evaluating Different Reponses to the Challenges of Platform Power. *Journal of Digital Media & Policy, 12*(2), 231–246.

Flew, T., Martin, F., & Suzor N. (2019). Internet Regulation as Media Policy: Rethinking the Question of Digital Communication Platform Governance. *Journal of Digital Media and Policy, 10*(1), 33–50.

Global Alliance for Responsible Media (GARM). (2021). Uncommon Industry Collaboration to Improve Digital Safety. World Federation of Advertisers. About GARM. https://wfanet.org/leadership/garm/about-garm

Gawer, A. (2021). Digital Platforms and Ecosystems: Remarks on the Dominant Organizational Forms of the Digital Age. *Innovation*. https://doi.org/10.1080/14479338.2021.1965888.

Gilbert, R. J. 2021. Separation: A Cure for Abuse of Platform Dominance? *Information Economics and Policy, 54*. https://doi.org/10.1016/j.infoecopol.2020.100876.

Gillespie, T. (2017). Governance of and by Platforms. In J. Burgess, T. Poell, & A. Marwick (eds.), *SAGE Handbook of Social Media* (pp. 254–278). Los Angeles: Sage.

Gorwa, R. (2019). The Platform Governance Triangle: Conceptualising the Informal Regulation of Online Content. *Internet Policy Review, 8*(2). https://doi.org/10.14763/2019.2.1407.

Hall, K. (2020, April–June). Public Penitence: Facebook and the Performance of Apology. *Social Media + Society*, 1–10.

Halpern, D. (2015). *Inside the Nudge Unit*. London: Penguin.

Helmond, A. (2015). The Platformization of the Web: Making Web Data Platform Ready. *Social Media + Society, 1*(2), 1–11. https://hdl.handle.net/11245/1.490538.

Kasperkevic, J. (2021, April 9). Chinese Antitrust 2.0: Why Is China Going After Its Big Tech? *Promarket: The Publication of the Stigler Center at the University of Chicago Booth School of Business*. https://promarket.org/2021/04/09/chinese-antitrust-exceptionalism-enforcement-trade-alibaba-zhang/.

Khan, L. (2018). Sources of Tech Platform Power. *Georgetown Law Technology Review, 2*(2), 325–334.

Koop, C., & Lodge, M. (2017). What Is Regulation? An Interdisciplinary Concept Analysis. *Regulation & Governance, 11*(1), 95–108.

Kukutai, T., & Taylor, J. (2016). *Indigenous Data Sovereignty: Toward an Agenda*. Canberra: ANU Press.

Latour, B. (2007). *Reassembling the Social: An Introduction to Actor-Network-Theory*. Oxford: Oxford University Press.

Lee, Y. (2019). Situation of Human Rights in Myanmar. Report of the Special Rapporteur of the Human Rights Council to the United Nations General Assembly, Seventy-fourth Session. August 30, 2019. A/74/342.

Lotz, A. (2021). *Media Disrupted: Surviving Pirates, Cannibals, and Streaming Wars*. Cambridge, MA: MIT Press.

Macron, E. (2018, November 12). *Speech by M. Emmanuel Macron, President of the Republic at the Internet Governance Forum*. https://www.elysee.fr/en/emmanuel-macron/2018/11/12/speech-by-m-emmanuel-macron-president-of-the-republic-at-the-internet-governance-forum.

McAfee, A., & Brynjolfsson, E. (2017). *Machine, Platform, Crowd: Harnessing Our Digital Future*. New York: W. W. Norton & Co.

Mejias, U. A., & Couldry, N. (2019). Datafication. *Internet Policy Review, 8*(4). https://doi.org/10.14763/2019.4.1428

Mueller, M. (2010). *Networks and States: The Global Politics of Internet Governance*. Cambridge, MA: MIT Press.

Mueller, M. (2017). *Will the Internet Fragment?: Sovereignty, Globalization and Cyberspace*. Cambridge: Polity.

Musiani, F., Cogburn, D. L., DeNardis, L., & Levinson, N. S. (eds.). (2016). *The Turn to Infrastructure in Internet Governance*. Basingstoke: Palgrave Macmillan.

Napoli, P. (2019). *Social Media and the Public Interest: Media Regulation in the Disinformation Age*. New York: Columbia University Press.

Napoli, P., & Caplan, R. (2017). Why Media Companies Insist They're not Media Companies, Why They're Wrong, and Why It Matters. *First Monday*, 22(5). https://doi.org/10.5210/fm.v22i5.7051

O'Connor, D., & Schruers, M. (2016). Against Platform Regulation. *The Internet, Policy & Politics Conference*. Oxford Internet Institute, University of Oxford. http://blogs.oii.ox.ac.uk/ipp-conference/2016/programme-2016/track-c-markets-and-labour/government-regulation-of-platforms/daniel-oco nnor-matthew-schruers-against.html.

Parker, G., Van Alstyne, M., & Sangeet, P. (2016). *Platform Revolution: How Networked Markets Are Transforming the Economy*. New York: W.W. Norton & Co.

Puppis, M. (2010). Media Governance: A New Concept for the Analysis of Media Policy and Regulation. *Communication, Culture and Critique*, 3(2): 134–149. https://doi.org/10.1111/j.1753-9137.2010.01063.x. https://doi.org/10.1111/j.1753-9137.2010.01063.x

Reynders, D. (2020). Fifth Evaluation of the Code of Conduct. Directorate-General for Justice and Consumers. June 2020. European Commission. https://ec.europa.eu/info/sites/default/files/codeofconduct_2020_fac tsheet_12.pdf.

Rogoff, K. (2018, July 11). Has Big Tech Gotten Too Big for Our Own Good? *Project Syndicate*. https://www.marketwatch.com/story/has-big-tech-gotten-too-big-for-our-own-good-2018-07-02.

Rossi, A. (2018). How the Snowden Revelations Saved the EU General Data Protection Regulation. *The International Spectator*, 53(4), 95–111. https://doi.org/10.1080/03932729.2018.1532705.

Satariano, A. (2021, October 6). Facebook Hearing Strengthens Calls for Regulation in Europe. *New York Times*. https://www.nytimes.com/2021/10/06/technology/facebook-european-union-regulation.html

Savin, A. (2017). Chapter 1: Internet Regulation in the European Union. *EU Internet Law*. Cheltenham: Edward Elgar.

Schechner, S. (2021, November 10). Google Loses Appeal of $2.8 Billion EU Shopping-Ads Fine. *The Wall Street Journal*. https://www.wsj.com/articles/google-loses-eu-shopping-ads-case-appeal-11636539480.

Schlesinger, P. (2020). After the Post-Public Sphere. *Media, Culture & Society*, 42(7–8), 1545–1563. https://doi.org/10.1177/0163443720948003.

Schewe, E. (2021, February 4). After the Capitol Riot, Who Will Govern Speech Online? *JSTOR Daily*. https://daily.jstor.org/after-the-capitol-riot-who-will-govern-speech-online/.

Sims, R. (2021). Digital Competition—Should the Australian Confrontation of Big Tech Serve as a Role Model for the EU? Interview Transcript. Stiftung

Neue Verantwortung (SNV) https://www.stiftung-nv.de/en/publication/transcript-digital-competition-should-australian-confrontation-big-tech-serve-role-model.

Stigler Center for the Study of the Economy and the State. (2019). *Stigler Committee on Digital Platforms: Final Report*. Chicago Booth. https://research.chicagobooth.edu/stigler/media/news/committee-on-digital-platfo rms-final-report.

Thaler, R. (2015). *Misbehaving: The Making of Behavioral Economics*. New York: W.W. Norton & Co.

The Economist. (2018, January 20). *The Techlash Against Amazon, Facebook and Google—And What They Can Do—A Memo to Big Tech*. https://www.economist.com/briefing/2018/01/20/the-techlash-aga inst-amazon-facebook-and-google-and-what-they-can-do.

Thompson, G. F. (2003). *Between Hierarchies and Markets: The Logic and Limits of Network Forms of Organization*. Oxford: Oxford University Press.

Tufecki, Z. (2018, April 6). Why Zuckerberg's 14-Year Apology Tour Hasn't Fixed Facebook. *WIRED*. https://www.wired.com/story/why-zuckerberg-15-year-apology-tour-hasnt-fixed-facebook/.

Van Dijck, J. (2021). Seeing the Forest for the Trees: Visualizing Platformization and Its Governance. *New Media & Society, 23*(9), 2801–2819. https://doi.org/10.1177/1461444820940293.

Van Dijck, J., De Winkel, T., & Schäfer, M. T. (2021). Deplatformization and the Governance of the Platform Ecosystem. *New Media & Society*. https://doi.org/10.1177/14614448211045662.

Warren, C. (2021, February 24). Diversity Hit Between the Eyes as old Media Pockets About 90% of Big Tech Cash. *Crikey*. https://www.crikey.com.au/2021/02/24/media-diversity-hit-old-media-big-tech-cash/.

Winseck, D. (2020). Vampire Squids, 'the Broken Internet' and Platform Regulation. *Journal of Digital Media & Policy, 11*(3), 241–282. https://doi.org/10.1386/jdmp_00025_1.

Wu, T. (2018). *The Curse of Bigness: Antitrust in the New Gilded Age*. Columbia Global Reports.

Zuckerberg, M., & Senate Commerce, Science and Transportation Committee. (2018, April 11). Transcript of Mark Zuckerberg's Senate hearing. *Washington Post*. https://www.washingtonpost.com/news/the-switch/wp/2018/04/10/transcript-of-mark-zuckerbergs-senate-hearing/.

CHAPTER 2

Can Journalism Survive in the Age of Platform Monopolies? Confronting Facebook's Negative Externalities

Victor Pickard

INTRODUCTION

Many of the world's print news media outlets today are facing existential threats from the collapse of their advertising revenue-based business model. Much of the blame for this decline has focused on the role of platforms such as Facebook and Google, which devour the lion's share of digital advertising revenue. The sustainability of journalism in general, and local news in particular, is increasingly threatened by this duopoly. In the U.S., the duopoly controls over 70% of the total online advertising market (including roughly 85% of all new U.S. digital advertising revenue growth), leaving only a pittance for news publishers (Kafka 2018; Myllylahti 2018). Meanwhile, institutions that provide actual quality news and information are being weakened by the loss of audiences and revenues to

V. Pickard (✉)
Annenberg School for Communication, University of Pennsylvania,
Philadelphia, PA, USA
e-mail: victor.pickard@asc.upenn.edu

© The Author(s) 2022
T. Flew and F. R. Martin (eds.), *Digital Platform Regulation*,
Palgrave Global Media Policy and Business,
https://doi.org/10.1007/978-3-030-95220-4_2

23

platforms at a time when democratic societies desperately need reliable journalism. Thus far, however, policy measures to rebalance this market power have been limited at best.

With its massive lobbying power, Facebook wields tremendous political-economic influence over not just information and communication systems themselves, but also the debates about how to regulate these systems. This regulatory capture helps explain how Facebook has been able to deflect responsibility for its actions for so long and maintain its posture as a neutral technology platform. While many argue that Facebook should be treated as a media company and held to relevant legal duties and obligations—as well as norms of social responsibility—Mark Zuckerberg has long refused to even acknowledge that Facebook is anything more than a technology company. While this problem deserves close public scrutiny, history shows us that expecting good corporate behavior simply by shaming information monopolies is a dubious proposition at best and arrangements for self-regulated "social responsibility" are often insufficient (Pickard 2015; Nurik 2021). Building on recent work that draws from historical lessons to argue for a "new social contract" (Pickard 2021) and a "reckoning" with the predictable threats to democracy posed by a lightly regulated, highly commercialized media system (Pickard 2022), my analysis in this essay moves beyond the critique of monopoly power to consider systemic solutions for sustaining digital journalism, especially public alternatives.

After addressing key debates around the harms that Facebook inflicts upon democratic societies, I discuss proposals for platform regulation that range from compelling platform companies to fund a journalism trust to reinventing a new public media system for the digital age. I conclude by focusing on more radical proposals for alternatives to the current profit-driven system, including public ownership. While much of this analysis centers on the U.S. political economy, it holds important implications for democratic nations around the world.

FACEBOOK'S NEGATIVE EXTERNALITIES

Given that media markets produce various externalities (Baker 2002), it is the role of government policy to manage them—to minimise the negative and maximize the positive externalities for the benefit of democratic society. As Facebook extracts profound wealth across the globe, it has generated tremendous negative externalities by mishandling users' data, abusing its market power, spreading dangerous misinformation and propaganda, and enabling interference in democratic elections in places such as the US and the Philippines (Vaidhyanathan 2018), and even playing role in facilitating ethnic cleansing in Myanmar (Stevenson 2018). As Facebook hurts democracy around the world and disserves its nearly 3 billion users—through mass surveillance, discrimination, and amplifying dis/misinformation and hate speech—it continues to shirk the democratic responsibilities that should automatically attend to any firm that controls such far-reaching communication infrastructures (Pickard 2020b). Given its record thus far, it is now abundantly clear that the firm has garnered far too much power and must be reined in, a concern reflected in chapters across this collection. Amid all this scrutiny, one area of harm is increasingly capturing the attention of policy analysts, critics, and scholars in recent years: Facebook's effects on news media and journalism (e.g., Bell and Owens 2017; Myllylahti 2018; Martin and Dwyer 2019; Meese and Hurcombe 2020; Napoli 2019; Pickard 2020a).

In various ways, Facebook's monopoly power corrupts the integrity of vitally important news and information systems. It acts as an algorithm-driven gatekeeper over a primary information source for its billions of users. In the U.S., where Americans increasingly access news through the platform (Gramlich 2021), Facebook's role in amplifying disinformation has drawn well-deserved scrutiny. Moreover, Facebook's Basics project has made it the sole portal to the internet for some countries, creating an unhealthy dependency. And, as noted earlier, Facebook and Google are siphoning most of the digital advertising revenue and starving the traditional media publishers that provide original news and information—the same struggling news organizations that these platforms expect to help fact-check against dis/misinformation (Kafka 2018). Journalism's financial future is increasingly threatened by the Facebook-Google duopoly. At the same time, Facebook in particular is accelerating the spread of dis/misinformation online.

Research consistently shows that commercial news organizations are increasingly relying on social media—especially Facebook—to reach audiences (Cornia et al. 2018), which has several troubling consequences. For starters, it incentivizes editors and journalists to make editorial decisions based on how news stories will likely perform on Facebook, thereby pandering to Facebook's algorithms and users' behavior. This exploitative and corruptive relationship pervades every aspect of journalistic labor and content. Facebook's position as the primary news portal to millions of readers forces precariously employed journalists to tailor their reporting according to what is essentially click-bait criteria.

Making matters worse, editors often reinforce this warped power relationship by constantly informing reporters how their work is faring on Facebook with real-time analytics flashing across their screens. Some newsrooms reportedly display wall-mounted data scoreboards provided by platforms such as Chartbeat, Parse.ly or Google Analytics that display social media metrics of specific stories and audience analytics (Petre 2021; Lamot and Van Aelst 2020; Fürst 2020). Moreover, the newsroom adoption of metrification underpins the ecosystem of companies like Chartbeat, which further intensifies and reinforces this logic (Martin and Dwyer 2019). These dynamics incentivize reporters to churn out controversial, trivial, and sensational content, that, in turn, encourages more people to engage with the Facebook platform for longer periods of time and, while under surveillance, producing more valuable information about themselves. Ultimately, this process generates more advertising revenue that mostly funnels back to Facebook instead of the news organizations and journalists who originally created the media content.

While public scrutiny of these unsavory practices continues to grow, much of it overlooks the *structural* roots of these problems, especially the commercial motives that accelerate it. Because its business model depends on user engagement and it profits handsomely from the attention paid to viral disinformation, hate speech, and other processes that cause social harm, Facebook is not incentivized to address these problems and is highly unlikely to do so to the extent that is necessary. This systemic failure underscores the need for structural reform. As this edited collection indicates, while there is no shortage of regulatory plans to address this long and growing list of negative externalities, thus far, the platforms have largely abdicated responsibility for taking action on them. One general reason the tech companies—and Facebook in particular—have been able to stave off regulatory interventions and greater accountability is their invocation of freedom of expression and other U.S. First Amendment rights. I interrogate these claims in the following section.

First Amendment Arguments

In the fall of 2019, Mark Zuckerberg made a highly publicized speech at Georgetown University (Kang and Isaac 2019), where he suggested that Facebook's first concern is to protect freedom of expression. Zuckerberg's speech was widely panned (and rightly so), but much of what he said reflects common contradictions of liberal democratic discourse. These ideological tensions create openings for someone like Zuckerberg to make outlandish claims that draw on common tropes but fail to withstand even the slightest scrutiny. From First Amendment absolutism to public sphere analysis, liberal/libertarian theories often ignore preexisting structural inequities and therefore often fail to acknowledge how some voices silence others. They conveniently presume a level playing field—an egalitarian "marketplace of ideas"—in which questions of power and exploitation have no purchase. And they typically conflate this marketplace of ideas with the capitalist market that directly and indirectly corrupts so many processes and practices within our communication and information systems.

According to this implicit "pay-to-play" logic, our media and communication systems function effectively if wealthy individuals and corporations can pay to be heard. Any attempt to confront this concentrated power—to create through regulation more opportunities for others to speak or to access information—is condemned as an illegitimate foray into the natural marketplace, seen as an attack on our core freedoms, and amounting to egregious censorship. But the market, itself an artificial creation, routinely censors and distorts speech and expression, especially when driven by advertising revenue (Baker 1994). For example, commercially-driven systems sort us into groups, surveil and target us with specific advertising, and ensure that some types of news information are not as readily available to certain audiences while privileging the access of others—especially those audiences most coveted by advertisers. Internet access itself is often determined by who can afford to pay for it, and according to specific corporate-friendly terms.

To pretend that the capitalist market is the best arbiter of permissible discourse is a core plank of "corporate libertarianism," defined by the notion that government has no legitimate role in media markets other than facilitating capital accumulation for a small number of elites (Pickard 2015). Of course, the state has *always* played a key role in designing information and communication systems and remains deeply involved in their governance; pretensions to the contrary are a libertarian fantasy. Nonetheless, it has been particularly challenging to have conversations about policy interventions in the U.S., where for many years discourse has been constrained by this corporate libertarian paradigm and First Amendment absolutism.

These inherent contradictions are rarely called out and provide cover for Zuckerberg to conflate his personal profit motives with the broader interests of democracy. The American media system's ideological foundation relies on an impoverished view of the First Amendment as dedicated to upholding negative liberties ("freedom from"—usually translated as "freedom from state interference") instead of positive liberties ("freedom for"), which might include protecting public access to a diverse and informative news media system (Berlin 1969). While the First Amendment encompasses *both* positive and negative liberties as essential to free expression in a democracy, the latter are easily captured by media corporations. These firms often exploit the libertarian qualities of negative liberties to use as a shield to deflect public interest regulations and public investments in alternative media infrastructures. It also naturalizes the unregulated market as the great defender of democratic discourse.

Public pressure can help steer monopolistic firms toward more responsible behavior for a time, though even that modicum of success is often contingent on a credible threat of regulatory intervention. And indeed, we have witnessed Facebook change course at times—for example, when it finally banned Trump in 2021. But the fact that it took an assault on the U.S. Capitol— an action that was in no small part organized on the platform (Mack et al. 2021)—to finally force Facebook's hand is very telling. In the next section, I look more closely at the political economic conditions that gird Facebook's position—and present opportunities for necessary structural reforms.

POLITICAL ECONOMIC ARGUMENTS

Many of the harms that Facebook externalizes to societies across the globe stem from its core business model, which relies on what is essentially a massive surveillance machine.[1] While many observers may have once viewed Facebook positively, the American public increasingly sees the company as a monopoly intent on doing whatever it takes to make as much money as possible (e.g., Reilly 2017). Moreover, like all monopolies, Facebook has shown that it will fight tooth and nail to retain that market power, even resorting to unsavory methods. A November 2018 *New York Times* story revealed that Facebook hired a disreputable public relations firm to smear adversaries with anti-Semitic conspiracy theories (Nicas and Rosenberg 2018). Pursuing profit to the detriment of democratic considerations, Facebook continually dodges efforts toward transparency and accountability.

At the same time, growing concerns about Facebook's unregulated power has engendered a rare bipartisan consensus that government must rein in platform monopolies. Until recently, the concept of regulating technology firms seemed unfathomable, but now even Republican policymakers—sometimes for ill-founded reasons such as the belief that Facebook is politically biased against conservatives—believe they have become so powerful that government must intervene. Even Zuckerberg, who is notoriously reluctant to take responsibility for causing social problems, has had to shift his rhetorical strategy to concede that perhaps Facebook should be subject to certain regulations (Isaac 2019). Facebook has continued to profess a pro-regulatory stance for several years.[2] On Valentine's Day 2021, Facebook even ran a full-page ad in the *New York Times* announcing that it supported "updated internet regulations." Of course, this begs the obvious question: *what kind* of regulation? Tamping down public criticism and responsibilities for content moderation—if it preserves profits—serves Facebook quite well.

Answering the question about what regulation should look like requires us to directly confront the impact of platform monopolies on journalism. Monopoly ownership is a broad structural threat to a healthy information system, affecting everything from control of internet access to the range of voices in our news media. Fortunately, a growing anti-monopoly movement in the US, as well as stronger stances toward platforms from many countries around the world, has offered some hope that these giants might be cut down to size. Indeed, in recent years have witnessed a growing clamor of antitrust initiatives, championed by politicians such as Senator Elizabeth Warren and advocacy groups such as the Open Markets Institute.

At the ideational level, this movement benefits from a growing consensus that something must be done to confront concentrated corporate power in general and the new tech monopolies in particular, coinciding with a growing "techlash" against Silicon Valley-based internet firms (*The Economist* 2018). A lively debate has emerged in recent years—mostly on the left but also including people from across the political spectrum—that champions what is sometimes referred to as the Jeffersonian or neo-Brandeisian approach, which emphasizes breaking up monopolies.

The neo-Brandeisian approach (named after Supreme Court Justice Louis Brandeis), which sees centralized control to be the most worrisome evil that must be prevented at all costs, focuses on breaking up concentrated market power and encouraging competition, primarily through antitrust measures (Wu 2018).

This framework underpins much of the U.S. anti-monopoly movement, whose main objective is to break up monopolies into smaller units along structural lines, thus creating a much more decentralized economic environment in which numerous firms compete for consumers. Anti-monopoly activists rightly identify the Chicago School of Economics as responsible for reorienting antitrust law toward what is known as the "consumer welfare standard," which emphasizes purported consumer benefits over public interest considerations such as unemployment and protecting small businesses. During the Reagan administration, this approach became the dominant paradigm, with the government willing to approve mergers so long as companies promised to keep prices low. Regulatory bodies exhibited less concern toward other well-known problems related to concentrated economic and political power, which led to highly concentrated industries exacting terrible social costs (Khan 2017).

There is much to admire in the anti-monopoly arguments. The platforms are, after all, simply too big. They have become vertically integrated monstrosities, wielding a degree of political power incompatible with a functioning democracy. Calls for "smashing them to bits" may sound quite appealing, even radical. But on closer examination we can see that this would not solve many of the information problems we face. While it is true that American antitrust enforcement has been overly lax for decades, leading to highly concentrated news and information industries, an over-emphasis on this strategy has drawbacks. Certainly, the ideal of maintaining robust competition among many small producers is noble— and the desire to break up vertically integrated monopolies, oligopolies, and cartels a legitimate and necessary objective.

A major limitation to the neo-Brandeisians' anti-monopoly approach, however, is its tendency to critique the size of monopolies and the lack of competition, rather than the commercial values that drive them toward perverse incentives. Such a critique tends to overlook structural questions about whether media systems should be governed by commercial motives and relationships in the first place. Simply reducing the size and multiplying the number of commercial outlets that depend on surveillant advertising, disseminating low-quality content, and undervaluing democratic concerns will likely not solve all our challenges. In other words, Facebook presents a capitalism problem, not just a monopoly problem.

Perhaps counter-intuitively, some progressive advocates argue for maintaining centralization. Part of this position rests on the notion that greater efficiencies stemming from scale and scope may create benefits for workers and consumers because large producers are easier to unionize and regulate. Within this regulated monopoly paradigm, big government can serve as a countervailing force against the excesses of big capital. The neo-Brandeisians, for their part, criticize this position as overly accommodationist, locking in and legitimating concentrated corporate power. The neo-Brandeisian notion that "big is bad"—or, as Brandeis himself referred to it, "the curse of bigness"—benefits from an intuitively resonant rhetoric of justice. Moreover, the desire to trust-bust monopolies has a populist appeal, connects with a rich history, and often presents itself as the radical—or at least the more progressive—option in policy debates. But in fact, the neo-Brandeisian approach is, in some ways, a deeply conservative position; it sees a fair and orderly market as the proper regulator of news media. In other words, it assumes that a highly capitalistic media system can serve democracy well, if only we managed it appropriately, especially via competitive markets.

It is becoming increasingly clear that antitrust is a necessary but insufficient intervention in designing a democratic communication and information system. While some advocates take this recognition to suggest the goal should be break-up *and* regulate (Kimmelman 2019), there is also a third way. What both the regulatory and antitrust approaches lack is a systemic critique of the market's failure to support public goods— that is, private firms' underproduction of the high-quality information that is fundamental for a democratic society to operate effectively. Unaccountable monopoly power is both a contributing factor to, and symptom of, this structural problem. If our ultimate goal is to create something different from the "surveillance capitalism" that drives so much of our digital news and information systems (Zuboff 2019; Foster and McChesney 2014), then it is clear that we need publicly-owned, democratic alternatives.

Much of the low-quality information that permeates through our news media system results from commercial pressures that privilege particular types of news coverage over others—not the malfeasance of a few bad journalists or news organizations. For example, Facebook designs its algorithms to encourage its users to engage with content on the platform primarily to sell targeted ads and drive corporate profits. As users, we are more likely to engage with material that has an emotional pull—for example, if something makes us angry or scares us. Hence, Facebook's algorithms reward content that fuels outrage—which mainstream news media produces by emphasizing social and political conflict. Consumer tracking and profiling encourages advertisers and news outlets to focus their efforts on narrowly tailored clickbait, regardless of a story's veracity. In the end, commercial logics and, specifically, the need to maximize profits via advertising revenue over all other concerns, drive our news and information systems, thereby enabling and amplifying misinformation. Low-quality information is not a defect of these systems, it is an essential feature.

The assumption that digital media somehow magically transcended these capitalistic imperatives was always an ideological assumption, not an empirical one—as even a cursory glance at the long historical record would indicate, from telephony to broadcast media. Indeed, by now the data are incontrovertible in demonstrating what happens when corporate monopolies dominate a highly commercialized information system. These systems are typically beset with predictable harms, hazards, negative externalities, and perverse incentives that might be good for business but are often very bad for democracy. The recurring unwillingness to see something so obvious is another reminder that if we do not understand the logics of a commercial media system—and the resultant effects of capitalism on news and information systems—we will always be taken by surprise by bad actors' bad behavior, and we will always ascribe this behavior to individuals—that of outliers and "bad apples"—instead of fundamental systemic flaws. Nonetheless, the never-ending quest for a self-regulatory fix continues apace, which I turn to next.

The Problem with Social Responsibility

The assumption that a social contract should guide corporate giants' operations is a recurring theme in policy history (Pickard 2021). Yet, unlike "natural monopolies" or public utilities of old, Facebook has avoided close regulatory oversight and shirked meaningful public interest requirements in exchange for the many benefits that society grants it. One possible approach toward finally establishing a new social contract might include a revamped "social responsibility" regime. On the surface, this may appear to be a positive development. And indeed, such an arrangement would likely be a marked improvement considering our current predicament. But democratic societies might consider cautionary tales and historical lessons before going too far down this path.

Social responsibility harkens back to earlier formations regarding regulation—or lack thereof—of the press. One key example is the U.S. Hutchins Commission—a blue ribbon panel tasked with defining press freedoms in the 1940s—which was famously tasked with the core question of "whether the giants should be slain or persuaded to be good." Ultimately, they decided that it was better to try to publicly pressure media firms into good behavior instead of aggressively regulate them. Although such experiments with media self-regulation have failed in many ways (Pickard 2015), this is precisely where we seem to be headed today with the Facebook Oversight Board and similar efforts that appear to show a self-reformed Facebook take on more responsibility. The reality, however, is that Facebook's political economic power—with all its attendant harms—remains intact.

At the same time, we are beginning to see interesting resolutions around the world that aim to recalibrate some of these power relationships. For example, the Australian approach to platform regulation recommended by the Australian Competition and Consumer Commission (ACCC 2019) has led to an instructive power struggle and gives us an opportunity to see a more aggressive standoff between platform monopolies and national governments. Generally speaking, this legislation forces platforms like Facebook and Google to compensate media companies for using their content. In actuality, however, many flaws and uncertainties compromise this plan, which has received critical scrutiny from academics and sundry critics, many of whom cast doubt on the efficacy of such policy interventions in saving local, independent journalism (Meese and Hurcombe 2020; Pickard 2019; Winseck 2020). Too often, this "platforms vs. publishers" frame emphasizes an aggrieved industry over the information needs of democratic societies. Nonetheless, this plan is serving as a default model that many countries are currently considering around the world.

While silver-bullet policy solutions are elusive, increased public scrutiny offers a fleeting opportunity to hold an international debate about what interventions are best suited to address informational deficits and social harms. Above all else, these problems necessitate *structural* reforms.

Shaming digital monopolies into good behavior or tweaking market incentives are, in the end, of limited utility. With platform monopolies accelerating a worldwide journalism crisis, a new social contract is required that, at the very least, includes platform monopolies paying into a global public media fund.

Taxing the Platforms to Fund Journalism

Although platform monopolies have not single-handedly caused the journalism crisis—overreliance on advertising revenue and structural shifts in the transition to digital formats are the primary causes—they have exacerbated the overall precarity facing the newspaper industry by defunding and compromising news content. Not only do these firms bear significant responsibility, they also command profound resources. But thus far, the platforms have funded only modest initiatives to support journalism, mostly bound up in optimizing news outlets' performance on Facebook (for an overview, see Pickard 2019, 2020a). Meanwhile, proposals have proliferated for more meaningful reforms that seek to redistribute revenues from the platforms to news publishers, with some seeking a more radical redress than others.

As noted earlier in discussing the ACCC plan, a general proposal that has gained much mainstream support—especially from politicians and publishers—is that the platforms should more fairly distribute their digital advertising revenues back to news media industries. At first glance, this seems fair and reasonable. But this proposal neglects the fact that there simply is not enough money in digital advertising to support the level of journalism that democratic societies require. In the U.S. alone, the newspaper industry has lost tens of billions of dollars since the early 2000s, predating the rise of Facebook. Thus, platform monopolies are responsible for only a percentage of such losses—even if the newspaper industry would argue that it is a very significant percentage (MaLoon 2019). A key concern with such schemes based on compensation to the aggrieved commercial news industry is the risk that these plans will disproportionately help incumbent big publishers who are themselves complicit in exacerbating the journalism crisis through consolidation and job cuts. Such restorationist proposals would arguably only reify the worst tendencies of commercial media, and likely make communication problems even worse, especially for disadvantaged communities (Pickard 2020a). Instead, privileging smaller independent, nonprofit organizations is arguably a better plan to ensure that such money actually goes toward sustainable, local journalism that societies need.

Another growing argument, one that I have also made (Pickard 2018, 2019, 2020a), is that the platforms should help fund the journalism that they are depriving of resources, but to direct that money towards funding *public* media. I propose that platforms pay a small percentage of their ample profits toward a journalism trust, which would generate hundreds of millions of dollars per year. The media reform organization Free Press (Karr and Aaron 2019) has similarly called for a digital advertising tax (which, of course, would be applied almost entirely toward the two big platforms), and the advocacy group Public Knowledge has called for a creating a "super fund" that the platforms pay into to help finance public service journalism (Stella 2020). Another proposal has called for establishing a $1 billion international public interest media fund to support investigative news organizations around the world, protecting them from violence and intimidation (Lalwani 2019). Similarly, the *Cairncross Review*, a detailed report on the future of British news media, called for a new institute to oversee direct funding for public-interest news outlets (Waterson 2019).

While all these proposed plans would be positive steps to varying degrees, ultimately such "offsets" do not strike at the core problem and could even be counterproductive for the long-term goal of taming and democratizing platform monopolies. In the following final section, I briefly discuss more radical and structural interventions.

Radical Imaginaries and Possibilities

Democratic societies faced with run-amok monopoly capitalism have three general tools at their disposal. First, they can break up monopolies and trust that a more competitive market will help tame destructive behavior. Second, they can regulate monopolistic firms and attempt to offset against social harms and negative externalities. Or third, they can try to create public alternatives that are not subject to the same market pressure and therefore, if designed and governed appropriately, can operate according to more socially beneficial logics. I have touched on the first two of these approaches but now will turn to the third option as I conclude this essay. While short-term reforms aimed at curbing Facebook's power and any efforts toward bolstering journalism should be applauded, we also must recognize that these are not, by and large, long-term, systemic solutions. In my view, only deep structural reform can assure democratic outcomes and more permanent solutions to the many problems we currently face.

To create the information and communication systems that democracy requires necessitates more radical, structural reform. While many such reforms—nationalizing the platforms, for example—are often cast out of bounds before the conversation even begins, and, moreover, the prospects of such radical futures are always remote, there are some promising signs afoot that suggest we can dare imagine more meaningful change. At the very least, it stands to argue, we should begin with attempting major structural reform before we fall back on less ambitious measures.

These more radical trajectories of change typically fall along several axes. One is the attempt to radicalize from within, especially among the tech workers themselves, who have been one the strongest vectors of political action against the platforms. For example, in recent years Google workers have made important political interventions, such as in 2018 when 20,000 employees staged a global walkout to protest sexual harassment. Given their key positions within the platform monopolies' larger power structures, tech workers could play an important role by organizing at multiple levels and democratizing these firms from the inside (Petcoff and Tarnoff 2021).

Another argument for reform—one grounded in mainstream economic theory and American history—is the notion that these platforms should be seen and treated as public utilities (Srnicek 2019; Schiller 2020; Muldoon 2020). If we start to move in that direction, we can easily imagine a host of new—and meaningful—public interest obligations (Pickard 2020b, 2021, 2022). While calls for renewed regulations from the broadcast era such as the Fairness Doctrine are likely implausible and unworkable for platforms, we could certainly argue for sensible protections and guardrails against the worst excesses (Pickard 2015; Napoli 2019). This might include "signal boosting" reliable information within Facebook news feeds and Google searches to increase its visibility in feeds and searches (Kornbluh and Goodman 2020).

Beyond regulatory measures, however, a more effective approach might be outright public ownership, including cooperative movements (see, for example, Hanna and Brennan 2020). Any move toward this direction would, over time, radically restructure labor relations and ownership structures, ultimately democratizing not only platform monopolies but entire sectors of our news and information systems. Some discussions around such proposal have emerged—for example, talk of a "public interest social media platform" in Australia and the notion that the *BBC* can be redesigned and expanded to compete with search engines and social media in the UK by presenting a non-commercial alternative. Even if, overall, these more radical reformist proposals are still in their infancy and remain mostly discursive, they are increasingly being taken seriously and may offer glimmers of hope for a more democratic future.

Ultimately, we must recall that media corporations are a social construct and their values, design principles, and relationships to individuals and communities are malleable according to social needs and public policies. As members of democratic societies, we have the power to change platform monopolies if we collectively decide to do so and make new policies for the greater good. It is up to us to decide that our media institutions must serve democracy first and foremost as opposed to mere profit imperatives. Our concerns should not be guided by the expectations of established industry players since this ongoing debate is not really about them, it is about us.

REFERENCES

Australian Competition and Consumer Commission (ACCC). 2019. *Digital platforms inquiry final report*. https://www.accc.gov.au/system/files/Digital%20platforms%20inquiry%20-%20final%20report.pdf

Baker, C. Edwin. 1994. *Advertising and a democratic press*. Princeton: Princeton University Press.

Baker, C. Edwin. 2002. *Media, markets, and democracy*. New York: Cambridge University Press.

Bell, Emily, and Taylor Owen. 2017. The Platform Press: How Silicon Valley reengineered journalism, *Columbia Journalism Review*, March 29. https://www.cjr.org/tow_center_reports/platform-press-how-silicon-valley-reengineered-journalism.php. Accessed July 5, 2021.

Berlin, Isaiah. 1969. Two concepts of liberty. In *Four essays on liberty*, ed. Henry Hardy, 118–172. Oxford: Oxford University Press.

Cornia, Alessio, Annika Sehl, David Levy, and Rasmus Kleis Nielsen. 2018. *Private sector news, social media distribution, and algorithm change*. Oxford: Reuters Institute. https://reutersinstitute.politics.ox.ac.uk/our-research/private-sector-news-social-media-distribution-and-algorithm-change. Accessed February 28, 2021.

Foster, John Bellamy, and Robert McChesney. 2014. Surveillance capitalism: Monopoly finance capital, the military-industrial complex, and the digital age. *Monthly Review* 66(3): 1–31.

Fürst, Silke. 2020. In the service of good journalism and audience interests? How audience metrics affect news quality. *Media and Communication* 8(3): 270–280.

Gramlich, John. 2021. 10 facts about Americans and Facebook. *Pew Research Center*, June 1. https://www.pewresearch.org/fact-tank/2021/06/01/facts-about-americans-and-facebook/. Accessed July 6, 2021.

Hanna, Thomas, and Michael Brennan. 2020. there's no solution to big tech without public ownership of tech companies. *Jacobin*. https://jacobinmag.com/2020/12/big-tech-public-ownership-surveillance-capitalism-platform-corporations

Isaac, Mike. 2019. Mark Zuckerberg's call to regulate Facebook, explained. *New York Times*, March 30. https://www.nytimes.com/2019/03/30/technology/mark-zuckerberg-facebook-regulation-explained.html. Accessed February 28, 2021.

Kafka, Peter. 2018. These two charts tell you everything you need to know about Google's and Facebook's domination of the ad business. *Recode*, February 13. https://www.recode.net/2018/2/13/17002918/google-facebook-advertising-domination-chart-moffettnathanson-michael-nathanson. Accessed February 28, 2021.

Kang, Cecilia, and Mike Isaac. 2019. Defiant Zuckerberg says Facebook won't police political speech. *New York Times*, October 21. https://www.nytimes.com/2019/10/17/business/zuckerberg-facebook-free-speech.html. Accessed February 28, 2021.

Karr, Timothy, and Craig Aaron. 2019. Beyond fixing Facebook: How the multibillion-dollar business behind online advertising could reinvent public media, revitalize journalism and strengthen democracy. *Free Press*, February 25. www.freepress.net/policy-library/beyond-fixing-facebook. Accessed February 28, 2021.

Khan, Lina. 2017. Amazon's antitrust paradox. *Yale Law Journal* 126: 710–805.

Kimmelman, Gene. 2019. To make the tech sector competitive, antitrust is only half the answer. *Public Knowledge*, February 22. https://www.publicknowledge.org/news-blog/blogs/to-make-the-tech-sector-competitive-antitrust-is-only-half-the-answer. Accessed February 28, 2021.

Kornbluh, Karen, and Ellen Goodman. 2020. *Safeguarding digital democracy: Digital Innovation and Democracy Initiative roadmap*. Washington, DC: German Marshall Fund. https://www.gmfus.org/publications/safeguarding-democracy-against-disinformation. Accessed February 28, 2021.

Lalwani, Nishant. 2019. A free press is the lifeblood of democracy—Journalists must not be silenced. *The Guardian*, July 5. www.theguardian.com/global-development/2019/jul/05/a free-press-is-the-lifeblood-of-democracy-journa lists-must-not-be-silenced. Accessed February 28, 2021.

Lamot, Kenza, and Peter Van Aelst. 2020. Beaten by chartbeat? An experimental study on the effect of real-time audience analytics on journalists' news judgment. *Journalism Studies* 21(4): 477–493.

Mack, David, Ryan Mac, and Ken Bensinger. 2021. "If They won't hear us, they will fear us": How the capitol assault was planned on Facebook. *Buzzfeed*, January 19. https://www.buzzfeednews.com/article/davidmack/how-us-cap itol-insurrection-organized-facebook

MaLoon, Michael. 2019. New study finds Google receives an estimated $4.7 billion in revenue from news publishers. *News Media Alliance*, June 10. https://www.newsmediaalliance.org/release-new-study-google-rev enue-from-news-publishers-content. Accessed February 28, 2021.

Martin, Fiona R., and Timothy Dwyer. 2019. *Sharing news online commendary cultures and social media news ecologies*. Palgrave Macmillan.

Meese, James, and Edward Hurcombe. 2020. Facebook, news media organisations and platform dependency: The institutional impacts of news distribution on social platforms. *New Media & Society*. https://doi.org/10.1177/146144 4820926472

Muldoon, James. 2020. Don't break up Facebook—Make It a public utility. *Jacobin*. https://jacobinmag.com/2020/12/facebook-big-tech-antitrust-soc ial-network-data/

Myllylahti, Merja. 2018. An attention economy trap? An empirical investigation into four news companies' Facebook traffic and social media revenue. *Journal of Media Business Studies* 15(4): 237–253.

Napoli, Philip. 2019. *Social media and the public interest: Media regulation in the disinformation age*. New York: Columbia University Press.

Nicas, Jack, and Matthew Rosenberg. 2018. A look inside the tactics of definers, Facebook's attack dog. *New York Times*, November 15. https://www.nytimes.com/2018/11/15/technology/facebook-definers-opposition-research.html. Accessed February 28, 2021.

Nurik, Chloé. 2021. Censored, commodified, and surveilled: How Facebook's self-regulatory governance harms marginalized communities. Dissertation, University of Pennsylvania.

Petcoff, Aaron, and Ben Tarnoff. 2021. Tech Workers at every level can organize to build power. *Jacobin*. https://jacobinmag.com/2021/02/tech-wor kers-organizing-class-position

Petre, Caitlin. 2021. *All the news that's fit to click: How metrics are transforming the work of journalists*. Princeton, NJ: Princeton University Press.

Pickard, Victor. 2015. *America's battle for media democracy: The triumph of corporate libertarianism and the future of media reform*. New York: Cambridge University Press.

Pickard, Victor. 2018. Break Facebook's power and renew journalism. *The Nation*, April 18. www.thenation.com/article/break-facebooks-power-and-renew-journalism. Accessed February 28, 2021.

Pickard, Victor. 2019. Public investments for global news. *Centre for International Governance Innovation*, October 28. https://www.cigionline.org/art icles/public-investments-global-news. Accessed February 28, 2021.

Pickard, Victor. 2020a. *Democracy without journalism? Confronting the misinformation society*. New York: Oxford University Press.

Pickard, Victor. 2020b. Restructuring democratic infrastructures: A policy approach to the journalism crisis. *Digital Journalism* 8(6): 704–719. https://doi.org/10.1080/21670811.2020.1733433

Pickard, Victor. 2021. A social contract for the new information monopolies: Historical lessons for the digital age. In *Regulating big tech: Policy responses to digital dominance*, ed. Damian Tambini and Martin Moore, 323–337. Oxford University Press.

Pickard, Victor. 2022. The great reckoning: Lessons from 1940s media policy battles. *Knight First Amendment Institute*. https://knightcolumbia.org/con tent/the-great-reckoning. Accessed March 4, 2022.

Reilly, Michael. 2017. Is Facebook targeting ads at sad teens? *MIT Technology Review*, May 1. https://www.technologyreview.com/2017/05/01/105987/ is-facebook-targeting-ads-at-sad-teens/. Accessed July 5, 2021.

Schiller, Dan. 2020. Reconstructing public utility networks: A program for action. *International Journal of Communication* 14: 4989–5000.

Srnicek, Nick. 2019. The only way to rein in big tech is to treat them as a public service. *The Guardian*. https://www.theguardian.com/commentis free/2019/apr/23/big-tech-google-facebook-unions-public-ownership

Stella, Shiva. 2020. Public knowledge: We need a 'Superfund for the Internet' to fight misinformation online. *Public Knowledge*, May 11. https://www.pub licknowledge.org/press-release/public-knowledge-we-need-a-superfund-for-the-internet-to-fight-misinformation-online. Accessed February 28, 2021.

Stevenson, Alexandra. 2018. Facebook admits it was used to incite violence in Myanmar. *New York Times*, November 6. https://www.nytimes.com/2018/ 11/06/technology/Myanmar-facebook.html. Accessed February 28, 2021.

The Economist. 2018. The techlash against Amazon, Facebook and Google—And what they can do. https://www.economist.com/briefing/2018/01/20/ the-techlash-against-amazon-facebook-and-google-and-what-they-can-do. Accessed July 5, 2021.

Vaidhyanathan, Siva. 2018. *Antisocial media: How Facebook disconnects us and undermines democracy*. New York: Oxford University Press.

Waterson, Jim. 2019. Public funds should be used to rescue local journalism, says report. *The Guardian*, February 11. www.theguardian.com/media/

2019/feb/11/public-funds-should-be-used-to-rescue-local-journalism-says-report. Accessed February 28, 2021.

Winseck, Dwayne. 2020. Vampire squids, 'the broken internet' and platform regulation. *Journal of Digital Media & Policy* 11(3): 241–282. https://doi.org/10.1386/jdmp_00025_1

Wu, Tim. 2018. *The curse of bigness: Antitrust in the new Gilded Age*. New York: Columbia Global Reports.

Zuboff, Shoshana. 2019. *The age of surveillance capitalism: The fight for a human future at the new frontier of power*. New York: Hachette/PublicAffairs.

CHAPTER 3

Platforms and the Press: Regulatory Interventions to Address an Imbalance of Power

Asa Royal and Philip M. Napoli

INTRODUCTION

A key question that any broad regulatory framework for digital platforms must address is what, if any, interventions are necessary to mediate the relationship between digital platforms and the news media. In many countries, the news media have long been subject to some form of government regulation and/or support, typically under the presumption that the cultivation and maintenance of an informed citizenry is essential to the effective functioning of the political process. But recently, digital platforms have emerged to establish a still-evolving, much-debated, and

A. Royal
DeWitt Wallace Center for Media & Democracy, Duke University, Durham, NC, USA
e-mail: asa.royal@duke.edu

P. M. Napoli (✉)
Sanford School of Public Policy, Duke University, Durham, NC, USA
e-mail: philip.napoli@duke.edu

© The Author(s) 2022
T. Flew and F. R. Martin (eds.), *Digital Platform Regulation*,
Palgrave Global Media Policy and Business,
https://doi.org/10.1007/978-3-030-95220-4_3

43

often unregulated position of prominence in national news ecosystems, serving as important intermediaries in the relationship between news organizations and their audiences.

Platforms' emergence into this role has had a wide range of well-documented, disruptive effects. Chief among these has been an acceleration of the unbundling of news and the further weakening of the revenue models associated with its production and distribution. Whereas would-be news readers were once forced to buy entire newspapers or visit ad-packed home pages to access stories, platforms now offer readers a smorgasbord of individual news stories they can sample (nearly) freely, empowering users to potentially reshape the nature of the news they receive (Martin and Dwyer 2019). Platforms have also tinted the windows of story discovery, guiding users' access to news with algorithmic content curation systems that favor emotionally charged and engagement-inducing content, veracity not necessarily withstanding (Ingram 2018; Rayson 2017). And though platforms' recommendation algorithms have garnered attention for their role in the propagation of disinformation, they have also created new incentive systems feeding directly into the editorial values that guide mainstream news organizations, promoting the publicisation, if not publication, of would-be viral content (Wang 2015).

Given these complexities, it is not surprising that the platform-press relationship has been rife with conflict. Nor is it surprising that policy-makers across many national contexts, concerned with maintaining the robust news ecosystems essential to democracy, have increasingly turned their attention to the relationship.

This chapter focuses on efforts by policymakers to mandate that digital communications platforms (including search engines and social networks) that host or show any news content compensate the content's publishers. The approach in this chapter is cross-national and comparative, focusing on three countries (France, Germany, and Australia) that have taken the most significant regulatory actions internationally in recent years, while at the same time noting actions that have taken place in other countries such as Belgium and Spain, and at the supranational level (e.g., the European Union). This chapter will consider not only the substance of the regulatory interventions that have been proposed and implemented, but also the political dynamics surrounding them and the critiques they have generated.

Case Study Overview

The three countries we chose as case studies have approached legislating platform-publisher relationships from different angles but still, as we detail below, have ended up acting out similar plays.

Those similarities owe much to the transitional trends that have affected publishers. For the past fifteen-odd years, consumer attention and ad revenue have ebbed away from the lucrative pages of newspapers and onto the platform-dominated internet. For almost as long, publishers and (American) platforms have engaged in a struggle over whether and how platforms, which link to publishers' content, should compensate publishers. The warring has focused on "snippets", short extracts of news articles often displayed alongside links; for example, on a Facebook post or in Google searches. Digital platforms claim they owe publishers nothing for using snippets; publishers disagree, but rely on digital platforms to distribute their content and thus have little say.

Their disagreements have taken on a familiar cadence: publishers sue (or national governments legislate against) platforms to get them to cough up, ostensibly over snippets; platforms battle the lawsuits or legislation and push publishers to drop their claims, sometimes offering compensatory sums in lieu of recognizing publishers' putative rights over snippets; if publishers do not comply, platforms play hardball, rallying the public against the legislation and dropping publisher-specific snippets from their services; publishers, facing traffic drops after their snippets have been dropped from platforms, give in, and the fight goes dormant until a few years later.

France

For the past fifteen years, French news publishers and (American) platforms have engaged in repeated iterations of the struggle just described. Only recently was the pattern broken. Following the French legislature's transduction of an E.U. directive establishing press copyrights over snippets, platforms (specifically Google) have begun inking the licensing deals with French news publishers that were once anathema to them.

French publishers' war with Google began in 2005, when the wire service Agence France-Presse (AFP) sued Google, alleging that Google News was illegally using AFP's content (Isbell 2010). As a wire service, AFP did not directly deliver news to readers, but instead offered it to news publishers under individual licensing agreements. When Google News crawled those publishers' sites, it found and freely displayed snippets of all content, including stories and images sourced from AFP. That, AFP claimed in its 2005 lawsuit, was illegal because Google had no license agreement for the AFP content. Google rejected AFP's claim, arguing that news, facts, and small bits of text like headlines and snippets could, not be copyrighted. AFP's suit, Google argued, threatened the freedom of information on the Internet (Isbell 2010).

According to AFP however, the organization was not suing Google over sharing intangible facts or language: it was suing Google for resharing and profiting from AFP's original photographs and stories, both of which, the company said, required significant effort and money to produce (*AGENCE FRANCE PRESSE v. GOOGLE INC.*, 2007). Google and AFP settled their case in 2007, and although financial terms of the deal were not released, Google ended up acquiring a license for AFP's content, just as it had earlier that year with the Associated Press (Auchard 2007).

By 2012, French news publishers were in the midst of a financial crisis afflicting news organizations worldwide. Despite government subsidies of €1.2 billion, not a single national newspaper was profitable (*The Economist* 2012). Amidst that backdrop, French news publishers revived AFP's argument—that Google ought to compensate them for its usage of snippets from their articles.

Siding with French publishers, President François Hollande threatened to adopt neighboring Germany's *leisttungschutzreicht* (LSR), a law creating additional copyrights for the news media, should Google not pay up. That move prompted threats from Google to completely delink French news media sites from Google search results (AFP 2012; Ternisien 2013). But in February of 2013, following a new and supposedly unrelated proposal by Hollande to tax Google over its data collection practices, the company settled with French publishers, paying out a lump sum of €60 million into a digital innovation fund (Pfanner 2013). This payout framework was notably different than the Google-AFP deal. No licensing agreements were struck, and despite significant support from the national government, French publishers received no acknowledgement of their "ownership" of snippets (Schmidt 2013).

Google had reached a similar outcome with Belgian publishers three months earlier. Settling a lawsuit over snippets and content caching, the company had paid an unspecified amount (reportedly €5 million, much of which was made up of purchases on Google's own ad platform) while unequivocally stating that it was not compensating news publishers for the right to use their content (Geerts 2012). In Google's persistent argument, previewing article snippets in Google Search was not illegal content appropriation, but an efficient improvement to the web, good for the end-user. And if previews deterred some searchers from visiting the snippeted site, Google theorized, those lost page views were more than made up for by the traffic its engine referred (Silva 2020a).

Despite the Google payouts, French publishers continued to struggle economically. Newspaper revenue in the country dropped by over a third from 2007 to 2017, with over two-thirds of that loss stemming from decreased advertising (Assouline 2019; Autorité de la concurrence 2020). Meanwhile, other EU countries tried and failed to funnel platform money to news publishers. In Spain, the government passed legislation that would force Google to pay publishers for displaying snippets. Google responded by shutting down Google News Spain, damaging Spanish news publishers' traffic (Athey et al. 2017). In Germany, the government approved an ancillary press copyright, after which Google turned off snippets for publications that would not sign free licensing deals, depressing their traffic numbers until they conceded two weeks later (Fels 2014).

Change in Europe did not arrive until 2019. After a court threw out Germany's LSR, the EU passed its Directive on Copyright in the Digital Single Market, Article 15 of which extended traditional press copyrights to include "neighboring" rights protecting snippets (*Council Directive 2019/790*, 2019), which could be wielded against large digital platforms. Notably, the EU principle of *subsidiarity* prevents the Union from enacting legislation unless the goals of the legislation cannot be achieved by individual member states acting alone (*Consolidated Version of the Treaty on European Union—TITLE 1, Article 5*, n.d.). Thus, the directive's passage demonstrated the European Parliament's belief that it would take a consolidated union rather than a few lone states to coerce platform compliance on copyrights.

Before the EU Directive passed, France, where licensing battles had begun nearly 20 years earlier, had already queued up press copyright legislation; and in July of 2019, three months after the directive's passage, the country ratified its new copyright law as a transduction of the directive (Assouline 2019; Piquard 2019). In response, as it had done in Germany and Spain before, Google refused to bargain with French news publishers, instead asking all of them for free licenses to use snippeted content. The company's line to publishers was essentially "We don't value your snippets enough to pay for them. Give them to us for free, or we'll drop them and you'll face the traffic crashes German holdouts did." Google's VP of News published a blog stating, in general terms, that the company would not pay publishers for people clicking on links, though the EU/French legislation dealt with the display of snippets, not the clicking of links (Gingras 2019).

French publishers sued Google, arguing the company was ignoring the spirit of the new law (as captured in its title, *Proposal for a law to create a neighboring right for the benefit of press agencies and press publishers*). The law, publishers claimed, explicitly stated it was designed to protect the press' financial investments in producing news, to uphold the public's interest in a free and pluralist news ecosystem (*Council Directive 2019/790*, 2019). The French Competition Authority agreed. In its decision, the Authority accused Google of using its dominant market position (the company controlled 93% of the national search market at the time) to dictate terms to publishers, who were reliant on the company's irreplaceable referral services. The Authority argued that by asking companies to give up content rights in exchange for those services and universally refusing negotiation, Google was abusing the economic dependence of others, a breach of European competition law (Autorité de la concurrence 2020). Moreover, the Authority noted, the EU directive article on press copyrights was written with the explicit purpose of shifting bargaining power away from platforms and to news publishers, a group the directive hailed as essential to information availability and democracy (*Council Directive 2019/790*, 2019)· The Authority's decision ordered Google to bargain in good faith with publishers over license compensation, and after a failed legal appeal by Google, was upheld by a French court (Rosemain 2020).

In November of 2020, a handful of French news publishers became the first in the world to sign government-mandated content licensing agreements with a digital platform—in this case, Google (Missoffe 2020; Rosemain 2021). By January of 2021, a major French press union representing nearly 300 titles announced it had reached a framework finalizing agreements with Google on behalf of nearly half its members, with more to come (L'Alliance Presse 2021).

These licensing agreements supposedly require Google to compensate news publishers for the company's use of content falling under the publishers' neighboring rights, though Google and the publishers disagree on whether that means Google is paying for snippets. APIG, the press union, has pointed out that Google only signed the licensing agreements after being sued under new EU/French laws (as the company admitted in a blog post) (Lomas 2021; Missoffe 2020) that established copyrights for snippets (Reda, n.d.). Therefore, the union says, the neighboring rights deals obviously cover snippets (Lomas 2021). A January 2021 statement by Google showed that the company officially begs to differ (Lomas 2021).

The deals are opaque, at least to the public. Individual news publishers will reportedly be paid pre calculated amounts on three-year contracts, but remuneration for their copyright licensing will be subsumed into payments from Google's News Showcase program, which compensates publishers for their "editorial expertise" and for allowing Google's customers beyond-the-paywall access to news content (Bender 2020; European Publishers Council 2020; Lomas 2021). Such legal legerdemain obscures how much Google is paying for what and explains how the company, even after signing licensing deals that stemmed from lawsuits over snippet appropriation, can still maintain that they are not compensating publishers for snippets.

Google has long maintained that Google Search is more of an information conduit—a platform—than a content service (Silva 2021b). The former designation, based on Section 230 of the U.S. Communications Decency Act, has more neutral connotations than the latter and, at least in the US, confers a number of legal protections (Kosseff 2019); as such, Google has clear incentives to maintain that it will not pay for content (Silva 2021b). Additionally, were Google to pay news publishers, other content creators might well come knocking for their own share.

As of June 2021, France is the only EU country that has implemented the press publisher section of the EU copyright directive (The International Association On the Digital Public Domain 2020). The supposedly mandatory deadline for doing so has, in fact, passed by, as EU member states wrestle with other sections of the directive. But if and when countries do transduce the new press laws, they may well walk the path paved by France and Google, wherein compensation for publishers' neighboring copyrights is subsumed into a larger program like Google's News Showcase, and publishers accept a much-needed money line in exchange for not pressing copyright issues.

GERMANY

In 2011, German Chancellor Angela Merkel delivered a speech to the Federal Association of German Newspaper Publishers (BDZV), calling for the government to pass new copyrights protecting press publishers in the digital world. Thus began a tussle with Google.

In trying to protect publishers, the government pursued a legalistic path that would update copyright laws to include a *leistungsschutzrecht*—or ancillary copyright—for news snippets. The LSR offered news publishers a one-year copyright term, during which they would have the exclusive right to license snippets for commercial use by search engines and similar services (Achtes Gesetz Zur Änderung Des Urheberrechtsgesetzes 2013). Under the LSR, if Google wished to display snippets from *Der Spiegel* stories in its search service, it would be up to the outlet to decide whether and how much Google should pay.

As Germany's intercession on behalf of news publishers came through copyright law, protests, unsurprisingly, came not just from platforms, but from advocates of information freedom like Wikimedia Deutschland and Creative Commons (Abrell, n.d.; *Unterstützer*, n.d.). Some critics decried the ambiguity of the law—it did not explicitly set a minimum length of content qualifying for copyright, nor did it clearly define the types of companies who would have to pay copyright fees (Kreutzer et al. 2011). Others claimed that meaningful news copyrights were encapsulated by existing copyright law and that the LSR, because it covered small bits of text, gave publishers a newfound and dangerous power to copyright facts and chunks of the German language (Max-Planck-Institut für Ibender 2013).

Google, the main target of the LSR, called the day of its passage a "black day for the Internet in Germany" ("Google lehnt Lizenzierungspflicht ab," 2012). A spokesman expressed the company's belief that economic partnerships were a better path forward than laws (a foreshadowing of Google's deal making in France eight years later) ("Google lehnt Lizenzierungspflicht ab," 2012).

As the LSR wound its way through the Bundestag, Google began its hardball routine. Before the bill passed, Google set up an anti-LSR website titled "Defend Your Net" and urged users to lobby politicians in opposition to the law (Google 2012). Then, after the LSR became law, Google asked German news publishers to sign a waiver allowing the company to freely use snippets extracted from their articles. A group of publishers (notable exceptions included *Der Spiegel*) refused and banded together under a collection society, VG Media (now Corint Media) to demand payment for snippeting. Google, in turn, decided that the cost of paying for snippets was not worth their value, or that it could bully news publishers into submission; when the LSR became active, the company stopped showing snippets from the rebellious publishers in its search engine and news service. Traffic to the news publishers of VG Media subsequently plummeted. Axel Springer, a German media giant, reported a loss of 40% of Google Search traffic and 80% of Google News traffic to its properties during the snippet abstention period (Fels 2014). And so two weeks after revoking Google's free snippeting rights, Springer and most of the publications under VG Media relented, allowing Google to resume snippeting (ten Wolde and Auchard 2014).

For years after Google's triumph, the LSR existed in a phantom state, challenged by lawsuits, and ineffective in helping publishers claim licensing fees. That lasted until 2019, when the LSR was felled by the European Court of Justice over a technicality: Germany had not notified the EU of the legislation, rendering it illegal (*VG Media Gesellschaft zur Verwertung der Urheber- und Leistungsschutzrechte von Medienunternehmen mbH v Google LLC, successor in law to Google Inc.*, 2019).

In 2018, the LSR's principles were formally adopted by the EU in its Directive on Copyright in the Digital Single Market (*Council Directive 2019/790*, 2019). A short while later, preempting German transduction of the EU directive into national law, Google began dispersing funds to news publishers. In June of 2020, the company signed licensing agreements totaling $300M with publishers in Germany (as well as publishers in Australia and Brazil), heralding in its press release, a quote from the head of Spiegel Group, one of the publishers that had originally allowed Google to freely use snippets and did not join the bargaining collective (Bender 2020). Then, in October of 2020, Google launched News Showcase, another product that would allow publishers to license content to the company. Its announcement once more hailed three German publications that had declined to join VG Media and bargain against Google (Pichai 2020; *VG Media*, n.d.).

Google News Showcase was in fact open to all publishers, but some, including Germany's Axel Springer, stayed away, arguing that Google's largesse came with contractual strings and might be yanked away if news publishers participated in legal claims under the EU directive (European Publishers Council 2020). Indeed, Axel Springer signed a content licensing deal with Facebook in May of 2021, only on the explicit grounds that the deal would not cover future copyright claims (Wienker 2021). Most news publishers have neither the clout nor the capital of Axel Springer. Just as they did in France, publishers in Germany have signed on in large number to Facebook and Google's news licensing schemes, eschewing fights about the legal position of snippets to gain access to cash infusions.

Australia

In 2019, when the Australian government first announced it would draft legislation to support news publishers in their fight to claim compensation from digital platforms, it had already witnessed the nearly decade-long travails of European governments making similar attempts.

Perhaps because of the troubles European governments had faced, the Australian government sharply diverged from their model of legislation. Rather than pursuing a press copyright that would give publishers a narrow avenue to extract payment from digital platforms, the Australian government took a broader approach. Citing antitrust and public interest philosophies, it brought forth a mandatory news bargaining code governing platform-publisher relationships, the final version of which included provisions forcing platforms to, among other things:

1. Pay news publishers for the right to link to or show snippets from news stories at a rate subject to final offer arbitration.
2. Turn over data to news publishers about platform users' interactions with their content.
3. Notify news publishers in advance about platform algorithm updates that might affect the ranking or display of their content (*Treasury Laws Amendment (News Media and Digital Platforms Mandatory Bargaining Code) Bill 2020*, 2020).

The introduction of Australia's mandatory bargaining code followed over a decade of complaints by domestic publishers that digital platforms (which Australian media baron Robert Murdoch once called "content kleptomaniacs") effectively steal content by displaying snippets and previews of news without paying licensing fees (Dawber 2011; Sarno 2009).

Three years prior to the bargaining code's introduction, then-treasurer (now Prime Minister) Scott Morrison announced an inquiry into digital platforms' effect on competition in media and advertising. One goal of the inquiry was to investigate the impact of digital platforms on the public supply of news and journalism (Australian Competition and Consumer Commission 2019). The Digital Platforms Inquiry report found that platforms' dominant position as information distributors had given them significant bargaining power over the news companies whose content they displayed. The report resolved that given the news industry's vital importance to democracy, the major platforms should each establish a code of conduct addressing that bargaining imbalance (Australian Competition and Consumer Commission 2019). After giving platforms and publishers about a year to negotiate over what such a code might look like, the Australian government announced in 2020 that it had lost confidence in their talks (Australian Competition and Consumer Commission 2021). Thus, the mandatory bargaining code was born.

In the past, Google had responded to legislative forays in Germany and Spain by pressuring news publishers to ignore the legislation and give the company free license to display snippets. The Australian legislation attempted to tie Google's hands. Under the bargaining code, platforms could no longer just eliminate certain publishers' snippets to avoid paying them, because links, too, would require a license to display (*Treasury Laws Amendment (News Media and Digital Platforms Mandatory Bargaining Code) Bill 2020*, 2020). And because the code contained a non-discrimination clause, links from certain publications could not simply be delisted (Australian Competition and Consumer Commission 2021): the legislation rolled must-pay and must-carry provisions together. At any rate, it would be difficult for Google to run a search engine without linking to news content, especially given the term's loose definition. For the company, the legislation presented a binary nuclear option: bargain on new terms with publishers or leave the Australian search market.

Australian government intervention in the platform-publisher relationship came during a challenging time for the country's news publishers. Following global trends, Australian newspaper revenue had collapsed over the prior two decades. From 2002 to 2018, Australian papers lost 23% of their subscription revenue and 87% of their classified ad revenue, the latter of which, lost to online specialist sites like *Carsales* (an auto advertising website) and seek (a job advertising site), made up 92% of papers' overall revenue decline (AlphaBeta Australia 2020). Between 2008 and 2018, the number of local and regional papers in the country declined by 15%, a loss of 106 outlets, the closures of which left 21 local government areas with no local or regional paper. The fall of papers, noted the Australian government, had been deleterious to democracy. In 2018, the ACCC (Australian Consumer and Competition Commission) found that in surviving papers, reporting on local government and local courts had dropped by 26 and 40% respectively (Australian Competition and Consumer Commission 2019).

The government's abrupt timing in imposing the mandatory bargaining code—abrogating the voluntary negotiating period—was brought on by a period of especially acute pain for Australian news publishers (Crowe 2020). Reduced advertising due to the Covid-19 pandemic led many Australian news organizations to close or cut back operations, even as the pandemic and an unusually active bushfire season spiked domestic demand for news (Helliker 2020). News Corp, Australia's most prominent publisher, announced that 36 of its papers in the country would shut down, and another 76 would fully migrate online. Australian Community Media, the country's largest owner of regional and rural publications, cut back operations at 77 papers over the course of the pandemic (The Public Interest Journalism Initiative, n.d.).

Strikingly, as Australian newspapers' revenue declined, international demand for news increased. From 2013 to 2018, the global number of online news subscriptions rose 307%, growing by 26 million and eclipsing the 0.5% (3 million) decrease in print subscriptions (AlphaBeta Australia 2020).

Growing demand for news has coincided with massive growth for platforms like Facebook and Google, which have expanded to capture a dominant share of the digital advertising revenues in Australia (AlphaBeta Australia 2020; Hunter and Samios 2020). Some have linked the rise of platforms to the fall of newspapers (Cantwell 2020; Kang 2020; Stoller 2019; Sullivan 2021). Data, however show that newspaper display ad revenue rose approximately 6% from 2002 to 2018 (AlphaBeta Australia 2020). Of course, as more advertising dollars migrate online, the proportion controlled by the largest platforms dwarfs the share that news organizations are able to capture. Further, as submissions to the ACCC's Digital Platform Inquiry note, many news organizations contend that platform growth has been built on the backs of publishers, with the platforms' dominant market position enabling them to bully news organizations into ceding free content. As that bargaining imbalance was the impetus for government action, payouts from platforms to publishers under the bargaining code are meant as compensation for platforms' display of publishers' news content.

Of course, in redistributing power to news organizations, the government took it away from platforms companies, which responded with umbrage. Immediately after the draft bargaining code's announcement, Google and Facebook both cited internal figures suggesting that news content is responsible for only small portions of their traffic and revenue (Barrett and Kaye 2020; Silva 2020b), and, moreover, that platforms delivered outsized referral benefits to the news media (Facebook 2020; Google 2021). Facebook announced that should the legislation pass, it would block all news sharing on the platform (Cheik-Hussein 2020). Google launched an offensive in the public sphere, displaying pop-up ads on its services that asked users and content creators to lobby the Australian government against the bargaining code (Zhou 2020). The company also made repeated reference to its contributions to the Australian economy, stressing that it provides a platform to 1.3 million domestic businesses, contributes "$53 billion in benefits" to the Australian economy, and "supports 116,000 jobs across the country" (Google 2021; Silva 2019, 2021b).

The platforms were not the only stakeholders that were critical of the bargaining code. A number of analysts raised concerns that by allowing compensation to be determined via negotiations between platforms and individual news organizations, the Australian system would ultimately favor large, established, national news organizations relative to smaller, local, or independent news organizations with even less bargaining power (Hui and Tripti 2021). Many small publishers are not eligible for compensation from the platforms under the code (Samios 2020). Outside the media, others raised concerns that allowing entities to charge others for the ability to display a hyperlink wound fundamentally undermine the web (McGuirk and Chan 2021; Visentin 2021a).

As debate over the bargaining code continued, in June of 2020, Google announced the launch of News Showcase, a licensing program through which the company would pay news publishers in exchange for editorial curation and offering Google users access to normally paywalled articles (Bender 2020). Citing its example in France, Google offered to compensate Australian publishers through News Showcase (in place of arbitrated bargaining), but Australian officials and publishers expressed skepticism (Samios and Visentin 2021). Nonetheless, in late January of 2021, Google accelerated the timeline for rolling out News Showcase in Australia (Samios and Visentin 2021). At the same time, Google threatened to wholly remove Google Search from the Australian market should the bargaining code not change (Cave 2021a; Silva 2021a). According to the company, its breaking points included the mandate of compensation decided by final offer arbitration, restrictions on linking, and the requirement to detail algorithm updates in advance to news publishers (Silva 2021b).

Facebook took its objections one step further, blocking the viewing and sharing of all news links for Australian users in February 2021 (Cherney 2021). This Facebook news blackout was short-lived, lasting about a week, until the Australian government made some concessions in the terms of the bargaining code, including giving platforms more time to negotiate with publishers and also allowing platforms to avoid the bargaining code if they struck enough deals with individual news publishers (Australian Competition and Consumer Commission 2021; Isaac and Cave 2021; Meade et al. 2021). And so, concurrently with the concessions and the lifting of the blackout, Google and Facebook began signing licensing deals with Australian publishers (News Corp 2021; Visentin 2021b). The legislation officially passed immediately after (Boom 2021).

The eight day Facebook news blackout served as a catalyst for a more intensive analysis of the role of large digital platforms in news ecosystems. On the one hand, the bulk of the news blocked from being posted and shared on Facebook was still directly accessible online, which highlights the importance of not conflating large digital platforms with the broader Internet. On the other hand, research showed immediate and substantial drops in traffic to Australian news sites as a result of the blackout (Purtill 2021). Most news organizations are not in a financial position to absorb such traffic (and associated revenue) losses for any prolonged period of time. However, there is evidence that news consumers respond to the loss of news sources on Facebook by accessing them directly, such that initial traffic losses can be overcome over time (Mercer 2021; Napoli 2019).

Also, as some analysts have pointed out, for some segments of the population, accessing news outlets directly is a costlier proposition than doing so through a platform. Specifically, a small (generally lower-income) subset of the population relies primarily on mobile devices for their

Internet access and often has pre-paid rate plans in which Facebook access is cheaper than general web access (Chanel 2021). For this segment of the population, the Facebook news blackout may have deprived them of their most affordable mechanism for accessing news online.

The other issue that the Facebook blackout brought to the forefront is how dominant digital platforms define news. When Facebook instituted its news blackout, analysts quickly noted that the platform's operational definition of news was both expansive and idiosyncratic (Cave 2021b). In addition to news organizations, Facebook blocked the posts of state health departments and emergency and weather services. Posts for some political candidates were blocked, as were those of some unions and nonprofit groups working with victims of poverty and domestic violence. But despite this expansive (and difficult to justify) definition of news, posts by conspiracy theorists and anti-vaccine groups remained up (Cave 2021b).

And while some of the pages that Facebook blocked were quickly restored, others took over week, prompting questions from some of the company's critics as to whether the initial expansiveness of the blackout was an intentional show of force in their negotiations with the Australian government (Cave 2021b). Such accusations, if true, are troubling; as is the alternate explanation—that the company is that ill-equipped to effectively define a news organization. The end result, in any case, has been additional fuel to the fire of concerns about the massive gatekeeping power wielded by a select few digital platforms (see, e.g., Scola 2021).

Conclusion

The ripple effects of what has taken place in France and Australia have been widespread, with policymakers in the United States and Canada vowing to follow Australia's lead (Espinoza and Barker 2021; Klar 2021; Ljunggren 2021), even in light of how disruptive and contentious the situation became. Should the government efforts described in this chapter's three case studies migrate to other countries, then it would seem that the question of whether dominant digital platforms *should* compensate news organizations will be one of the past, and the questions left will be about *how* that will transpire.

As of June 2021, the Australian Treasurer has signaled that as long as Facebook and Google continue making side payments to publishers, the other requirements of the News Media and Digital Platforms Mandatory Bargaining Code can be forgotten (Isaac and Cave 2021). Likewise, the French government has been silent on arguments about the copyright-worthy sanctity of snippets. There, it seems that so long as Google compensates publishers, the government will overlook whether or not the company is specifically paying for (or says it is paying for) snippets. For all intents and purposes, both laws have become elaborate mechanisms for extracting money from platforms to deliver to the struggling news media. But even if the laws are meant to be throwaway tools for leveraging money, they still shape its dispersion amongst outlets, potentially distorting the laws' public interest goals.

As noted above, one primary critique of the Australian model has been that it favors large, national news organizations over local and/or independent outlets, a side-effect of provisions which see Google and Facebook paying out money in part according to traffic (Missoffe 2021) and require the companies to simply sign "enough" deals before being freed of the code (Meade et al. 2021). Another critique holds that the Australian and French models prop up old-school journalism outfits, disincentivising evolution of the press (Ingram and Jarvis 2021), and perhaps also undermining access to diverse viewpoints.

But other approaches, potentially better at supporting diverse, locally-oriented sources of news, exist. One such proposal suggests taxing platforms and placing funds in an endowment tasked with equitably supporting local, independent, and non-commercial journalism, or supporting a network of local fact-checking organizations (see, e.g., Karr and Aaron 2019; *Superfund for the Internet Proposal Summary*, n.d.). Another proposal advocates using national infrastructure funds to provide media vouchers to citizens, thus allotting support to news media along grassroots preference lines (Waldman 2021). But it remains to be seen whether these alternative approaches will gain traction in subsequent national contexts given the prominence of competition law as a contemporary instrument in media regulation. Much will depend, in all likelihood, on how the situations in Australia and Europe play out in the short term.

A key goal in any approach should—from a freedom of the press standpoint—be to minimize to the extent possible the role that governments play in determining which news organizations receive platform funding and how the available funds are distributed across them. An independent news media demands as much. But, as in so many aspects of platform governance, the platforms have not engendered confidence in terms of their ability to make decisions that are well-attuned to serving communities' information needs (Napoli 2019). Ultimately, as these case studies have illustrated, platforms possess tremendous leverage in their relationships with even the largest news organizations, and so, in the absence of an existential redefinition of the news media or a massive transition to a primarily non-commercial model of journalism (probably the preferred outcome in all of this; see Pickard 2019), government mandates of some type seem essential to assuring that at least some of the advertising revenues that platforms have diverted from the news ecosystem find their way to the news organizations that are so vital to an informed citizenry.

References

Abrell, B. (n.d.). Die Politik ist der Lobbyarbeit der Verlage aufgesessen. *Max-Planck Gesellschaft*. Retrieved December 21, 2020, from https://www.mpg.de/7483234/leistungsschutzrecht.

Achtes Gesetz zur Änderung des Urheberrechtsgesetzes. (2013). https://www.buzer.de/gesetz/10638/a181081.htm?m=/87f_UrhG.htm.

AFP. (2012, October 18). Google threatens to stop linking to French media sites. *France 24*. https://www.france24.com/en/20121018-google-threatens-stopping-linking-french-new-sites-media-law-content-france-usa.

AGENCE FRANCE PRESSE v. GOOGLE INC. (US District Court for the District of Columbia April 6, 2007). https://dockets.justia.com/docket/district-of-columbia/dcdce/1:2005cv00546/113951.

AlphaBeta Australia. (2020). *Australian Media Landscape Trends*. https://alphabeta.com/wp-content/uploads/2020/09/australian-media-landscape-report.pdf.

Assouline, D. (2019). *Proposition de loi tendant à créer un droit voisin au profit des agences de presse et des éditeurs de presse* (No. 243). http://www.senat.fr/rap/l18-243/l18-243.html.

Athey, S., Mobius, M. M., & Pál, J. (2017). *The impact of aggregators on internet news consumption* (SSRN Scholarly Paper ID 2897960). Social Science Research Network. https://papers.ssrn.com/abstract=2897960.

Auchard, E. (2007, April 7). AFP, Google news settle lawsuit over Google news. *Reuters*. https://www.reuters.com/article/us-google-afp-idUSN0728115420070407.

Australian Competition & Consumer Commission. (2019). *Digital platforms inquiry final report*. https://www.accc.gov.au/system/files/Digital%20platforms%20inquiry%20-%20final%20report.pdf.

Australian Competition & Consumer Commission. (2021, February 25). *News media bargaining code* [Text]. Australian Competition and Consumer Commission. https://www.accc.gov.au/focus-areas/digital-platforms/news-media-bargaining-code.

Autorité de la concurrence. (2020). *Décision n° 20-MC-01 du 9 avril 2020*. Autorité de la concurrence. https://www.autoritedelaconcurrence.fr/sites/default/files/integral_texts/2020-04/20mc01.pdf.

Barrett, J., & Kaye, B. (2020, September 1). Explainer: Facebook, Google battle Australia over proposed revenue-share law. *Reuters*. https://www.reuters.com/article/us-australia-media-facebook-explainer-idUSKBN25S3YZ.

Bender, B. (2020, June 25). A new licensing program to support the news industry. *The KeyWord*. https://blog.google/outreach-initiatives/google-news-initiative/licensing-program-support-news-industry-/.

Boom, D. V. (2021, February 24). Australia passes law forcing Google and Facebook to pay news publications. *CNET*. https://www.cnet.com/news/australia-passes-law-forcing-google-and-facebook-to-pay-news-publications/.

Cantwell, M. (2020). *Local journalism: America's most trusted news sources threatened*. U.S. Senate Committee on Commerce, Science, and Transportation. https://cdn.arstechnica.net/wp-content/uploads/2020/10/Local-Journalism-Report-10.26.20_430pm_Draft.pdf.

Cave, D. (2021a, January 22). An Australia with no Google? The bitter fight behind a drastic threat. *The New York Times*. https://www.nytimes.com/2021/01/22/business/australia-google-facebook-news-media.html.

Cave, D. (2021b, February 18). Facebook's new look in Australia: News and hospitals out, aliens still in. *The New York Times*. https://www.nytimes.com/2021/02/18/business/media/facebook-australia-news.html.

Chanel, S. (2021, February 18). Facebook's Australia ban threatens to leave Pacific without key news source. *The Guardian*. https://www.theguardian.com/world/2021/feb/19/facebooks-australia-ban-threatens-to-leave-pacific-without-key-news-source.

Cheik-Hussein, M. (2020, June 15). Facebook says it doesn't need news content for Australia—AdNews. *AdNews*. https://www.adnews.com.au/news/facebook-says-it-doesn-t-need-news-content-for-australia.

Cherney, M. (2021, February 18). Facebook surprises Australians with news blackout. *Wall Street Journal*. https://www.wsj.com/articles/the-day-facebook-went-dark-on-news-11613652431.

Consolidated version of the Treaty on European Union—TITLE 1, Article 5. (n.d.). [Text/html; charset=UTF-8]. Official Journal 115, 09/05/2008 P. 0018 - 0018; OPOCE. Retrieved January 27, 2021, from https://eur-lex.europa.eu/LexUriServ/LexUriServ.do?uri=CELEX:12008M005:EN:HTML.

Council directive 2019/790. (2019). EUR-Lex. https://eur-lex.europa.eu/eli/dir/2019/790/oj.

Crowe, D. (2020, April 19). "A level playing field": Digital giants will have to pay for news. *The Sydney Morning Herald*. https://www.smh.com.au/pol itics/federal/a-level-playing-field-digital-giants-will-have-to-pay-for-news-202 00419-p54l7q.html.

Dawber, A. (2011, October 23). Murdoch blasts search engine "kleptomaniacs." *The Independent*. https://www.independent.co.uk/news/media/online/mur doch-blasts-search-engine-kleptomaniacs-1800569.html.

Espinoza, J., & Barker, A. (2021, February 9). EU ready to follow Australia's lead on making Google and Facebook pay for news. *Los Angeles Times*. https://www.latimes.com/world-nation/story/2021-02-08/eu-australia-google-facebook-tech-news.

European Publishers Council. (2020, October 1). Google's news showcase challenges EU publisher's right. *EPC*. https://www.epceurope.eu/post/google-s-news-showcase-challenges-eu-publisher-s-right.

Facebook. (2020). *Facebook response to the Australian treasury laws amendment (news media and digital platforms mandatory bargaining code) bill 2020*. https://www.accc.gov.au/system/files/Facebook_0.pdf.

Fels, E. (2014, May 11). *Axel Springer concludes its data documentation: Major losses resulting from downgraded search notices on Google*. Axel Springer SE. https://www.axelspringer.com/en/press-releases/axel-springer-concludes-its-data-documentation-major-losses-resulting-from-downgraded-search-notices-on-google.

Geerts, T. (2012, December 12). Partnering with Belgian news publishers. *Google Europe Blog*. https://europe.googleblog.com/2012/12/partnering-with-bel gian-news-publishers.html.

Gingras, R. (2019, September 25). How Google invests in news. *Google*. https://blog.google/perspectives/richard-gingras/how-google-invests-news/.

Google. (2012, November 30). *Verteidige Dein Netz*. https://web.archive.org/web/20121130070556/http://www.google.de/campaigns/deinnetz/.

Google. (2021, January 22). 8 facts about Google and the news media bargaining code. *Google | The Keyword*. https://blog.google/around-the-globe/google-asia/australia/8-facts-about-google-and-news-media-bargaining-code/.

Google lehnt Lizenzierungspflicht ab. (2012, August 29). *Spiegel*. https://www.spiegel.de/netzwelt/netzpolitik/google-lehnt-lizenzierungspflicht-ab-a-852 775.html.

Helliker, J. (2020, April 14). *Australians turn to digital news to stay informed*. Nielsen. https://www.nielsen.com/au/en/press-releases/2020/australians-turn-to-digital-news-to-stay-informed.

Hui, M., & Tripti, L. (2021, February 19). In Australia news showdown, it's Facebook and Google vs News Corp. *Quartz*. https://qz.com/1974645/in-australia-news-showdown-its-facebook-and-google-vs-news-corp/.

Hunter, F., & Samios, Z. (2020, July 30). Tech giants' ad revenue overwhelms major Australian media companies. *The Sydney Morning Herald.* https://www.smh.com.au/politics/federal/tech-giants-ad-revenue-overwhelms-major-australian-media-companies-20200730-p55gzq.html.

Ingram, M. (2018, January 22). *Facebook's latest changes will probably make misinformation worse.* Columbia Journalism Review. https://web.archive.org/web/20210414025641/https://www.cjr.org/innovations/facebook-newsfeed-changes.php.

Ingram, M., & Jarvis, J. (2021, February 18). *Talking with Jeff Jarvis about forcing Big Tech to pay for news* [CJR Galley]. https://galley.cjr.org/public/conversations/-MTlTWstBcUp2j4sTaRE.

Isaac, M., & Cave, D. (2021, February 22). Facebook strikes deal to restore news sharing in Australia. *New York Times.* https://www.nytimes.com/2021/02/22/technology/facebook-australia-news.html?utm_source=sen dgrid&utm_medium=email&utm_campaign=Newsletters.

Isbell, K. A. (2010). *The rise of the news aggregator: Legal implications and best practices* (SSRN Scholarly Paper ID 1670339). Social Science Research Network. https://doi.org/10.2139/ssrn.1670339.

Kang, C. (2020, January 12). The decimation of local news has lawmakers crossing the aisle. *The New York Times.* https://www.nytimes.com/2020/01/12/technology/google-facebook-newspapers.html.

Karr, T., & Aaron, C. (2019). *Beyond-fixing-Facebook.* Free Press. https://www.freepress.net/sites/default/files/2019-02/Beyond-Fixing-Facebook-Final.pdf.

Klar, R. (2021, February 18). Congress faces news showdown with Facebook, Google. *The Hill.* https://thehill.com/policy/technology/539311-congress-faces-news-showdown-with-facebook-google.

Kosseff, J. (2019). *The twenty-six words that created the Internet.* Cornell University Press.

Kreutzer, T., Kierkegaard, S., & Ringnalda, A. (2011). Co-reach in IPR in new media workshop: CLSR co-hosts policy meeting on ISP liability. *Computer Law & Security Review, 27*(2), 213–219. https://doi.org/10.1016/j.clsr.2011.01.011.

L'Alliance Presse. (2021, January 21). L'@Alliance_Presse et @GoogleEn-France signent un accord relatif à l'utilisation des publications de presse en ligne. https://t.co/t2QEeBMwX3 [Tweet]. *@Alliance_Presse.* https://twitter.com/Alliance_Presse/status/1352153739315048448.

Ljunggren, D. (2021, February 18). Canada vows to be next country to go after Facebook to pay for news. *Reuters.* https://www.reuters.com/article/us-australia-media-facebook-canada/canada-vows-to-be-next-country-to-go-after-facebook-to-pay-for-news-idUSKBN2AI349.

Lomas, N. (2021, January 21). Google inks agreement in France on paying publishers for news reuse. *TechCrunch*. https://social.techcrunch.com/2021/01/21/google-inks-agreement-in-france-on-paying-publishers-for-news-reuse/.

Martin, F., & Dwyer, T. (2019). *Sharing news online: Commentary cultures and social media news ecologies*. Palgrave Macmillan. https://doi.org/10.1007/978-3-030-17906-9.

Max-Planck-Institut für Ibender. (2013, February 24). *Stellungnahme zum Gesetzesentwurf für eine Ergänzung des Urheberrechtsgesetzes durch ein Leistungsschutzrecht für Verleger*. https://web.archive.org/web/20130224082410/http://www.ip.mpg.de/files/pdf2/Stellungnahme_zum_Leistungsschutzrecht_fuer_Verleger.pdf.

McGuirk, R., & Chan, K. (2021, February 18). Australian media law raises questions about "pay for clicks". *AP NEWS*. https://apnews.com/article/business-europe-australia-media-journalism-771b10a4efd00d47a703655708f45c57.

Meade, A., Taylor, J., & Hurst, D. (2021, February 23). Facebook reverses Australia news ban after government makes media code amendments. *The Guardian*. http://www.theguardian.com/media/2021/feb/23/facebook-reverses-australia-news-ban-after-government-makes-media-code-amendments.

Mercer, C. (2021, February 24). *News site Stuff left Facebook. Seven months later, traffic is just fine and trust is higher*. Reuters Institute for the Study of Journalism. https://reutersinstitute.politics.ox.ac.uk/risj-review/news-site-stuff-left-facebook-seven-months-later-traffic-just-fine-and-trust-higher.

Missoffe, S. (2020, November 19). Un point sur nos avancées avec les éditeurs de presse en France. *Le Blog Officiel de Google France*. https://france.googleblog.com/2020/11/-droits-voisins.html.

Missoffe, S. (2021, January 21). L'Alliance de la Presse d'Information Générale et Google France signent un accord relatif à l'utilisation des publications de presse en ligne. *Le blog officiel de Google France*. https://france.googleblog.com/2021/01/APIG-Google.html.

Napoli, P. M. (2019). *Social media and the public interest: Media regulation in the disinformation age*. Columbia University Press.

News Corp. (2021, February 17). News Corp and Google agree to global partnership on news. *News Corp*. https://newscorp.com/2021/02/17/news-corp-and-google-agree-to-global-partnership-on-news/.

Pfanner, E. (2013, January 21). France proposes an Internet tax. *The New York Times*. https://www.nytimes.com/2013/01/21/business/global/21iht-datatax21.html.

Pichai, S. (2020, October 1). Our $1 billion investment in partnerships with news publishers. *Google | The Keyword*. https://blog.google/outreach-initia tives/google-news-initiative/google-news-showcase/.

Pickard, V. (2019). *Democracy without journalism?: Confronting the misinformation society*. Oxford University Press.

Piquard, A. (2019, July 23). « Droit voisin »: La France devient le premier pays à transposer la directive européenne. *Le Monde*. https://www.lemonde.fr/act ualite-medias/article/2019/07/23/droit-voisin-la-loi-ouvre-la-porte-a-une-dure-negociation-entre-medias-et-plates-formes_5492480_3236.html.

Purtill, J. (2021, February 24). Facebook's news ban "experiment" is almost over. Here's what we've learnt. *ABC News*. https://amp.abc.net.au/article/13183828?__twitter_impression=true&s=09.

Rayson, S. (2017, June 26). *We analyzed 100 million headlines. Here's what we learned (new research)*. Buzzsumo. https://web.archive.org/web/201708 06130553/http://buzzsumo.com/blog/most-shared-headlines-study.

Reda, J. (n.d.). Extra copyright for news sites ("Link tax"). *Julia Reda*. Retrieved March 2, 2021, from https://juliareda.eu/eu-copyright-reform/extra-copyri ght-for-news-sites/.

Rosemain, M. (2020, October 8). Google must talk to French publishers about paying for their content, court says. *Reuters*. https://www.reuters.com/art icle/us-france-google-copyrights-idUSKBN26T18J.

Rosemain, M. (2021, January 22). Google seals content payment deal with French news publishers. *Reuters*. https://www.reuters.com/article/us-france-google-publishers-idUSKBN29Q0SC.

Samios, Z. (2020, September 13). Youth, lifestyle websites fear ACCC code may "accidentally destroy media diversity." *Sydney Morning Herald*. https://www.smh.com.au/business/companies/youth-lifestyle-websites-fear-accc-code-may-accidentally-destroy-media-diversity-20200913-p55v43.html.

Samios, Z., & Visentin, L. (2021, January 26). Google backflips on news product launch amid political battle. *The Sydney Morning Herald*. https://www.smh.com.au/business/companies/google-backflips-on-news-product-launch-amid-political-battle-20210126-p56wyc.html.

Sarno, D. (2009, December 2). Murdoch accuses Google of news "theft." *Los Angeles Times*. https://www.latimes.com/archives/la-xpm-2009-dec-02-la-fi-news-google2-2009dec02-story.html.

Schmidt, E. (2013, February 1). Google creates €60m digital publishing innovation fund to support transformative French digital publishing initiatives. *Google*. https://blog.google/outreach-initiatives/google-news-initiative/goo gle-creates-60m-digital-publishing/.

Scola, N. (2021, February 19). Facebook just handed its critics in Washington a lot more ammunition. *Politico*. https://www.politico.com/news/2021/02/19/facebook-news-australia-google-470323.

Silva, M. (2019, September 17). Google on the ACCC digital platforms inquiry. *Google*. https://blog.google/around-the-globe/google-asia/australia/response-accc-final-report/.

Silva, M. (2020a, May 3). Responding to the revised publisher code process in Australia. *Official Google Australia Blog*. https://australia.googleblog.com/2020/05/responding-to-revised-publisher-code.html.

Silva, M. (2020b, May 31). Official Google Australia Blog: A fact-based discussion about news online. *Google Australia Blog*. https://australia.googleblog.com/2020/05/a-fact-based-discussion-about-news.html.

Silva, M. (2021a, January 12). Submissions to the draft News Media Bargaining Code show significant concern. *Google | The Keyword*. https://blog.google/around-the-globe/google-asia/australia/draft-news-media-bargaining-code-submissions/.

Silva, M. (2021b, January 22). Mel Silva's opening statement to the senate economics committee inquiry. *Google*. https://blog.google/around-the-globe/google-asia/australia/mel-silvas-opening-statement/.

Stoller, M. (2019, October 17). Tech companies are destroying democracy and the free press. *The New York Times*. https://www.nytimes.com/2019/10/17/opinion/tech-monopoly-democracy-journalism.html.

Sullivan, M. (2021, February 4). Perspective | These local newspapers say Facebook and Google are killing them. Now they're fighting back. *Washington Post*. https://www.washingtonpost.com/lifestyle/media/west-virginia-google-facebook-newspaper-lawsuit/2021/02/03/797631dc-657d-11eb-8468-21bc48f07fe5_story.html.

Superfund for the Internet proposal summary. (n.d.). Public knowledge. Retrieved February 25, 2021, from https://www.publicknowledge.org/superfund-for-the-internet-proposal-summary/.

ten Wolde, H., & Auchard, E. (2014, November 5). Germany's top publisher bows to Google in news licensing row. *Reuters*. https://www.reuters.com/article/us-google-axel-sprngr-idUSKBN0IP1YT20141105.

Ternisien, X. (2013, January 18). Les négociations entre Google et les éditeurs de presse s'enlisent. *Le Monde*. https://www.lemonde.fr/actualite-medias/article/2013/01/18/google-propose-50-millions-d-euros-a-la-presse-francaise_1819172_3236.html.

The Economist. (2012, November 10). Taxing times. *The Economist*. https://www.economist.com/international/2012/11/10/taxing-times.

The International Association on the Digital Public Domain. (2020, December 7). DSM Directive implementation update: Six months to go and no end in sight. *International Communia Association*. https://www.communia-association.org/2020/12/07/dsm-directive-implementation-update-six-months-go-no-end-sight/.

The Public Interest Journalism Initiative. (n.d.). *The Australian newsroom mapping project*. Public Interest Journalism Initiative. Retrieved December 2, 2020, from https://anmp.piji.com.au/.

Treasury laws amendment (news media and digital platforms mandatory bargaining code) bill 2020, Parliament of Australia (2020). https://parlinfo.aph.gov.au/parlInfo/search/display/display.w3p;query=Id%3A%22legislation%2Fbills%2Fr6652_first-reps%2F0000%22;rec=0.

Unterstützer. (n.d.). IGEL - Initiative gegen ein Leistungsschutzrecht. Retrieved January 4, 2021, from https://leistungsschutzrecht.info/unterstuetzer.

VG Media Gesellschaft zur Verwertung der Urheber- und Leistungsschutzrechte von Medienunternehmen mbH v Google LLC, successor in law to Google Inc., C-299/17 (Landgericht Berlin September 12, 2019). http://curia.europa.eu/juris/document/document.jsf?docid=208982&mode=req&pageIndex=1&dir=&occ=first&part=1&text=&doclang=EN&cid=17634789.

VG Media. (n.d.). VG Media. Retrieved December 18, 2020, from https://www.vg-media.de/en/right-holders.html.

Visentin, L. (2021a, January 19). "Fundamental challenge to free and open internet": Google VP Vint Cerf slams media code. *The Sydney Morning Herald*. https://www.smh.com.au/politics/federal/fundamental-challenge-to-free-and-open-internet-google-vp-vint-cerf-slams-media-code-20210119-p56v57.html.

Visentin, L. (2021b, February 23). Facebook to restore Australian news content after media bargaining code amendments. *Sunday Morning Herald*. https://www.smh.com.au/politics/federal/government-agrees-to-last-minute-amendments-to-media-code-20210222-p574kc.html.

Waldman, S. (2021, May 28). Opinion: Why local news should be included in the infrastructure bill. *Poynter*. https://www.poynter.org/business-work/2021/is-local-news-really-infrastructure/.

Wang, S. (2015, August 13). The New York Times built a slack bot to help decide which stories to post to social media. *Nieman Lab*. https://www.niemanlab.org/2015/08/the-new-york-times-built-a-slack-bot-to-help-decide-which-stories-to-post-to-social-media/.

Wienker, M. (2021, May 17). Axel Springer and Facebook agree on global cooperation. *Axel Springer SE*. https://www.axelspringer.com/en/press-releases/axel-springer-and-facebook-agree-on-global-cooperation.

Zhou, N. (2020, August 17). Google's open letter to Australians about news code contains "misinformation", ACCC says. *The Guardian*. http://www.theguardian.com/technology/2020/aug/17/google-open-letter-australia-news-media-bargaining-code-free-services-risk-contains-misinformation-accc-says.

CHAPTER 4

EU Digital Services Act: The White Hope of Intermediary Regulation

Amélie P. Heldt

Online Harms as the Linchpin of Intermediary Regulation

Intermediaries, mostly social media platforms, were at first been perceived as enablers of free speech online and as facilitators of a certain democratization of the public discourse (Tucker et al. 2017). Behind this appearance, their architecture and their algorithmic recommender systems have soon led to problems with illegal and harmful content (Gillespie 2014, p. 175; O'Callaghan et al. 2015). Indeed, critics soon identified that many intermediaries did not act against the dissemination of hate crime as well as non-criminal but harmful hate speech (Citron 2014). Neither did they prevent the spread of mis- and disinformation (Schulz 2019). Instead, their business model allegedly facilitates political micro-targeting and dark ads and amplifies conspiracy ideologies (Zarouali et al. 2020).

A. P. Heldt (✉)
Leibniz Institute for Media Research | Hans-Bredow Institut (HBI), Hamburg, Germany
e-mail: a.heldt@leibniz-hbi.de

© The Author(s) 2022
T. Flew and F. R. Martin (eds.), *Digital Platform Regulation*,
Palgrave Global Media Policy and Business,
https://doi.org/10.1007/978-3-030-95220-4_4

69

Until now, the primary law for intermediary regulation in the EU has been the E-Commerce-Directive. Under Art. 14 and 15 E-Commerce-Directive, intermediaries have no obligation to monitor user-generated content. They benefit from a liability exemption as long as they have no knowledge of illegal activities and act promptly upon notification. So far, this safe harbor regime protected intermediaries from regulation specifically targeting content moderation, and it substantially shaped the EU's digital market. All the more so because this has unleashed synergy effects with a similar law in the U.S., Sec. 230 of the Communication Decency Act, and created a de facto transatlantic market for platforms with user-generated content.

However, for the past four years, an amendment of this directive became an obvious priority due to the sequence of events. Since the first reports on the alleged voter manipulation via Facebook for the UK Brexiteer campaign, EU Member States respectively experienced the adverse effects of online speech harms (e.g., Germany with hate speech against refugees; France disinformation during the 2017 elections). Moreover, the self-regulatory efforts of platforms against online harms were considered neither efficient nor satisfactory by lawmakers. Consequently, single Member States pressed ahead and adopted laws targeting specific online harms. As the probably most discussed example, Germany passed the Network Enforcement Act (NetzDG), which forces platforms provide users with a complaint procedure for unlawful content (under German criminal law) and remove 'manifestly unlawful content' within 24 hours. Adopted in summer 2017, the NetzDG was an (explicit) reaction to self-regulatory initiatives' lack of efficiency.[1] Although this law and its implementation are highly criticized (Citron 2017; Kaye 2018), the call for more effective regulation against harmful online communication and subsequently limiting the platforms' power over free speech has become louder. France passed a law against information manipulation during election campaigns and introduced a new form of interim injunction. Furthermore, France also adopted a law against hate crime (Loi Avia), but the Constitutional Council overturned it for violations

---

[1] Speech by the then Federal Minister of Justice and Consumer Protection, Heiko Maas, on the bill to improve law enforcement in social networks (Network Enforcement Act) before the German Bundestag in Berlin on June 30, 2017, retrieved from https://www.bundesregierung.de/breg-de/service/bulletin/rede-des-bundesministers-der-justiz-und-fuer-verbraucherschutz-heiko-maas--793138.

of the proportionality test. Similarly, Austria adopted a Communication Platform Act (KoPlG). If more Member States followed the lead, the chances of a fragmentation of intermediary regulation within the EU would have increased, which partially explains the EU's eagerness to develop a common proposal (Cornils 2020, p. 77). The E-Commerce-Directive's provisions regarding intermediary liability in place were no longer considered sufficient and adequate (De Streel et al. 2020, p. 57).

Genesis of the DSA

In October 2019, the then-candidate for President of the EU Commission, Ursula von der Leyen, mentioned the Digital Services Act as a means to 'upgrade liability and safety rules for digital platforms' in her political agenda.[2] She also underlined the need to 'tackle issues such as disinformation and online hate messages' to protect democracies.

A leaked note in December 2019 revealed that the Commission considered the E-Commerce-Directive 'outdated' and that it needed to be replaced by a more comprehensive set of rules for digital services (Fanta and Rudl 2019). Regarding content moderation, the leaked note proposed to make uniform rules for the removal of illegal content binding across the EU and possibly include harmful (not necessarily unlawful) content. On a more technical side, the authors suggested maintaining the ban on general content monitoring in Art. 15 E-Com-Dir but re-consider special provisions for filter technologies.

The lawmaking process started in 2020 and is still ongoing. So far, it can be described as relatively speedy and as 'the biggest update of digital regulations for around two decades' (Lomas 2020). Several committees within the EU Parliament produced meaningful reports and developed recommendations.[3] Finally, the Commission presented its first 'Proposal

---

[2] Von der Leyen, U. (2019). *A Union that strives for more—My agenda for Europe*, retrieved from https://ec.europa.eu/info/sites/info/files/political-guidelines-next-commission_en_0.pdf.

[3] The JURI committee proposed standards and procedures for content moderation, and guaranteed access to remedies; as well as the establishment of a European Agency tasked with monitoring and enforcing compliance. The IMCO report called on the COM to propose concrete legislative measures including notice-and-action mechanisms; as well as a central regulatory authority for oversight and compliance; transparency requirements for advertising, nudging etc. The LIBE report also proposed the creation of an independent EU body to exercise effective oversight.

for a Regulation of the European Parliament and of the Council on a Single Market for Digital Services (Digital Services Act) and amending Directive 2000/31/EC' on December 15, 2020 (hereinafter DSA). This first proposal will serve as the basis for further deliberation and is, therefore, at the center of this paper. According to the Commission's proposal, the DSA ought to counteract the risks and problems that have arisen for both individuals and society as a whole from the use of information intermediaries, against the dependence of the economy and society on single providers, and the power of these providers over public discourse. Its goal is not to 'break' platforms but rather to constitute a common European rulebook to increase legal certainty for companies in the Digital Single Market, and, subsequently, better protect fundamental freedoms.

The EU Commission's Proposal

The DSA's application scope expands from mere hosting service (based on Art. 14 E-Com-Dir) to a more nuanced definition of addressees. According to Art. 2 (f) DSA, intermediary services include mere conduits, caching services, and hosting services. Art. 2 (h) defines online platforms as 'a provider of hosting service which, at the request of a recipient of the service, stores and disseminates to the public information'. According to Art. 25 (1) 'platforms which provide their services to a number of average monthly active recipients of the service in the Union equal to or higher than 45 million' are considered Very Large Online Platforms (VLOPs). Under Art. 16 DSA micro and small enterprises are excluded from the scope of application. By doing so, the Commission keeps its initial classification of intermediaries as neutral infrastructure providers laid out in Art. 14 and 15 E-Com-Dir but, at the same time, follows a gradual approach.

Enforcing National Laws

The DSA does not include an obligation for platforms to proactively review user content. Instead, Art. 7 DSA maintains the duty for the Member States to 'not impose a general obligation on providers monitoring obligation' (Art. 15 E-Com-Dir). The liability privilege remains as long as the platforms have no knowledge of illegal content. The decision to maintain this regime is probably due to the high risk of negative consequences for both the companies and the users' fundamental rights. The DSA proposal stipulates more exceptions, such as the obligation to act 'upon the receipt of an order to act against a specific item of illegal content, issued by the relevant national judicial or administrative authorities' 'without undue delay' (Art. 8 (1) DSA). Intermediaries are expected to deliver an immediate response expected, but the DSA does not spell out a concrete timeframe. However, it does include an obligation to act against users who regularly upload illegal content (Art. 20 DSA) and to report 'serious' crimes involving a threat to the life or safety of persons (Art. 21 DSA).

Most importantly, the DSA provides rules for the moderation of illegal content solely (Art. 2 (p) DSA), not for content that does not violate a legal prohibition. It leaves at the service's discretion whether to implement the measures for the enforcement of their respective content rules (community guidelines/standards). According to Art. 2 (g) DSA 'illegal content' means 'any information, which, in itself or by its reference to an activity, including the sale of products or provision of services, is not in compliance with Union law or the law of a Member State, irrespective of the precise subject matter or nature of that law'. Under Art. 14 DSA, providers of hosting services have to 'put mechanisms in place to allow any individual or entity to notify them of the presence on their service of specific items of information that the individual or entity considers to be illegal content.' If providers choose to remove or block content, they have to inform the user who posted the content and state the reasons for their decisions (Art. 15 DSA). Moreover, according to Art. 17 DSA, providers of online platforms need to provide users with an internal complaint-handling system.

Oversight and Enforcement

To monitor the addressees' compliance with the new rules and possibly enforce them, the DSA introduces two new oversight institutions: Digital Services Coordinators at the national level, and the Board for Digital Services at the EU level. These new public agencies would have specific supervisory rights with regard to the DSA—something the committee reports by the EU Parliament have been strongly advocating for.

Under Art. 38 (2) DSA each Member State shall designate a Digital Services Coordinator (hereinafter DSC) responsible for 'all matters relating to application and enforcement' of the DSA. For supervision, investigation, and enforcement, the DSC shall have special rights awarded by the DSA and common to all Member States. Moreover, they will have the authority to impose fines, to impose measures against a service's management, and, as ultima ratio, to decide over the interruption of a service if the DSC identifies repeated infringements (Art. 41 DSA). To allow for a harmonized approach within the EU, the DSCs shall cooperate with each other and with other competent authorities. The DSA lays the cornerstone for this new authority (Art. 39 DSA) but leaves any further development of the task at the Members States' discretion. States that already adopted a similar law could, for instance, merge the already existing competent authority at the national level with the DSC.

The DSCs will cooperate within an independent group and form the European Board for Digital Services (hereinafter the Board). The Board shall serve as an advisory body to the DSCs and the EU Commission (Art. 47 DSA) and form a superordinate structure intended to serve the purpose of better consultation and more effective application of the new rules. It will essentially facilitate the better coordination of supervision activities by the DSCs. Also, the Board will receive its special supervision rights for VLOPs. Under Art. 50 DSA, the enhanced supervision aims at avoiding systemic risks originating from the size of VLOPs and their subsequent influence on the public sphere. Altogether, the EU Commission, the Board, and the DSC have a wide range of measures at their disposal to enforce the rules set in the DSA. Additional interventions by the EU Commission in Art. 51, 58, and 59 DSA stipulate an active role for the DSC and the Board. This leads to a distribution of supervisory rights among different institutions in proceedings against VLOPs. Thereby, the imposition of the most severe sanctions is not only at the mercy of one competent authority.

Tools to Enhance Transparency and Accountability

Beyond concrete rules against the spread of hate speech and illegal content, European lawmakers also considered the need for more transparency about the intermediaries' activities and ways to possibly hold them accountable. Both aspects are essential for a better understanding of how intermediaries generally function and how they apply the new rules. At the individual level, users are the first beneficiaries of procedural guarantees regarding content moderation. According to the current proposal, their right to complain against illegal content and better understand corporate content moderation procedures are at the core of this regulation (Art. 12 (1) DSA).[4] This transparency right for users affected by content restrictions is concomitant to operational terms. According to Art. 12 (2) DSA, providers of intermediary services 'shall act in a diligent, objective and proportionate manner' when applying content restrictions on users. This includes a duty to respect the 'applicable fundamental rights of the recipients of the service as enshrined in the Charter.' Previously, the E-Com-Dir mentioned the importance of freedom of expression in its preamble. Intermediaries were expected to provide their 'information society services' in light of Art. 10 ECHR. The provisions, however, did not explicitly mention the ECHR.[5] This explicit obligation for intermediaries to take the fundamental rights enshrined in the EU Charter of Fundamental Rights into account is quite a novelty. It illustrates that lawmakers see the responsibility that should come along with the potential influence of intermediaries.

This leads us to the question of transparency at the corporate level: According to Art. 13 DSA, intermediaries will have to publish transparency reports at least once a year. Required information includes the number of orders received from MS, the number of notices submitted per Art. 14 DSA, content moderation activities engaged at the provider's initiative, and the number of complaints received in compliance with Art. 17 DSA. The reporting obligation increases in relation to the type and the size of the service. Under Art. 23, 33 DSA online platforms and VLOPs have to disclose additional information than mere intermediary services. VLOPs additionally have to provide an annual risk assessment (Art. 26 DSA), focusing on the usage of their services to disseminate illegal content, negative effects for fundamental rights arising out of their services, and the 'intentional manipulation of their service.' The latter is another novelty in terms of platform regulation: VLOPs are asked to assess their negative effect on the protection of public goods and, among others, their 'foreseeable impact related to electoral processes and public safety.' These obligations are paired with an annual independent audit at their own expense (Art. 28).

All in all, such reports can inform the public about the policies and practices of services that are heavily used all over the world but quite

---

[4] The right to access personal data under GDPR will not be affected; the rights can be cumulative.

[5] The EU Charter of Fundamental Rights did not exist when the E-Com-Dir was adopted.

opaque to most users so far. The effects could, therefore, not be limited to the EU but potentially inform stakeholders worldwide. Both types of instruments, at the individual and at the corporate level, can serve as information sources for complaints (Art. 43 DSA) and are, therefore, serving not only transparency but also accountability.

Interim Conclusion

At first look, the DSA proposal submitted by the EU Commission is more than a mere update of the E-Com-Dir. Under the new provisions, information and data would no longer be perceived only from the perspective of goods and markets. Instead, the DSA could become a human-rights-infused regulation (Llansó 2020) because not only does it explicitly mention the Charter of Fundamental Rights, it also builds in the values of the Charter in the provisions themselves. One fear (preceding the proposal) was that it would change the liability rules and force platforms to introduce "pro-active" measures against illegal content (Fanta and Rudl 2019), but following the heated discussion on upload-filter in the DSM-Directive (Heldt 2019), the Commission refrained from implementing such obligation in the DSA.

It is also worth noticing that the DSA is part of a larger package including, the Digital Market Act and the Democracy Action Plan. The latter is of particular interest for the questions regarding content moderation and fundamental rights in democratic societies. The EU Democracy Action Plan ought 'to ensure that citizens are able to participate in the democratic system through informed decision-making free from unlawful interference and manipulation.' With regard to the role of online platforms, the DAP includes six objectives (section 4.2):

1. monitoring the impact of disinformation and the effectiveness of platforms' policies;
2. supporting adequate visibility of reliable information of public interest and maintaining a plurality of views;
3. reducing the monetization of disinformation linked to sponsored content;
4. stepping up fact-checking;
5. developing appropriate measures to limit the artificial amplification of disinformation campaigns; and
6. ensuring an effective data disclosure for research on disinformation.

Ideally, the rules proposed in the DSA would serve as means to achieve the objectives set in the DAP. This interplay should be kept in mind when evaluating the single-out measures.

POTENTIAL FRICTIONS

Formal Matters

As mentioned in the history of the DSA, the setting is complicated due to pre-existing regulations by the Member States and matters of competency at the EU level. Generally, the EU is competent for the single market's realization (Art. 26 TFEU). According to Art. 114 TFEU, the EU Parliament, and the Council adopt legislation to harmonize the rules necessary to build and ensure the functioning of the single market. The EU does however, not have the legislative competency for criminal law. Hence, the definition of illegal content is left at the Member States' discretion. In the course of the DAP, the EU Commission plans to propose an amendment to Art. 83 TFEU 'to cover hate crime and hate speech, including online hate speech' in 2021 (On the European Democracy Action Plan 2020, p. 10). According to Art. 83 TFEU, the EU legislators can set 'minimum rules concerning the definition of criminal offences and sanctions in the areas of particularly serious crime with a cross-border dimension.' Such offences are thereby considered criminal and punishable in all Members States. The Commission's goal is to substantially enhance the protection of citizens and journalists and, therefore, to hamper further coarsening and polarisation within the public debate.

One should also carefully examine the necessity of new measures in the light of potential risks at the individual and collective level. In light of the developments in recent years, the legislator clearly needed to address contemporary issues of the digital sphere. One can measure the estimated need for harmonization at the EU level by the form chosen to legislate. Indeed, the DSA proposal comes as a regulation, which means that it will be applicable to all jurisdictions within the EU without any transposition legislation by the Members States. (As opposed to its predecessor, the E-Commerce-Directive.) This form limits the ability of Member States to deviate and to potentially dilute certain rules. The Commission seems to follow the GDPR's path (Wagner and Janssen 2021) based on the idea that a regulation will be more suitable for such a cross-border topic as the Digital Single Market than a directive.

Collision Risk

The DSA might collide with already existing laws. It remains unclear if the DSA should replace the Member States' existing laws like the German NetzDG or be considered supplementary to the DSA. The question is actually twofold. The DSA and pre-existing laws could either collide/diverge, or; they could respectively address aspects not mentioned by the other one. In the latter case, I believe that the DSA will serve as a minimum standard, allowing the Member States to adopt additional laws as long as it does not hollow out the DSA. This has to do with the principle of subsidiarity within the EU regarding the Member States' sovereignty. Even more so, because the DSA would not just contain new rules for the Digital Single Market but also overlaps with criminal law and media law. In cases where the DSA and national regulation could potentially contain contrary or very different rules for the same issue, the DSA would prevail. According to the precedence principle by the CJEU in the case Costa v Enel (1964), if a national rule is contrary to a European provision, Member States' authorities must apply the European provision.[6] National law is neither rescinded nor repealed, but its binding force is suspended. It is undisputed that this principle is indispensable for the functioning of European integration as a community based on the rule of law (Haltern 2020, p. 818). At the same time, Member States try to preserve a relevant influence over legislation as much as possible. The DSA could become yet another example of a tug of war between Brussels and national regulators.

Countering the Consolidation of Power Structures

The DSA's primary goal is to equilibrate the power structures in the digital economy, hence, to even out the dominant position of certain intermediaries over their users and their competitors (note, it is not an anti-trust law). Of course, the "big players," large companies from Silicon Valley, are in the "first line" because they developed a significant influence over the market by gathering data. Since the rise of the social web in the early 2000s, social media platforms have become a relevant communicative infrastructure. For most parts, they were the only

---

[6] Judgment of the Court of Justice of the European Communities of 15 July 1964. Flaminio Costa v E.N.E.L. Case 6/64.

arbiters of permissible expressions within their network (Suzor 2019) and have subsequently gained considerable power over their users' media diet. The modular setting of social media platforms (Schulz and Dreyer 2020, p. 31) makes it difficult to regulate them, that is, to use already existing frameworks or categories. Lawmakers are therefore compelled to conceive new regulatory approaches. After a long period of the self-regulation regime under Art. 14 E-Com-Dir (Buiten et al. 2020, p. 145), the EU decided to focus on stricter rules (Scott et al. 2020). That is why the DSA aims to strengthen the users' rights to be better informed, appeal certain decisions, and lodge complaints—regardless of the country and the laws restricting speech.

Does this approach really strengthen the platforms' power over online speech? One could argue that they will in the future still be the ones deciding over the removal of content, its distribution, and algorithmic recommendation systems. The new rules could perhaps consolidate their dominant position over the public discourse because it gives them more legitimacy. Two things can be said against this hypothesis. First, selection, prioritization, and recommendation are inherent to the service users look for: intermediaries provide this exact service, and users see "only" a selection of content. Second, the safeguards provided by the DSA on different levels will challenge the platforms in an unprecedented way. It might not be exhaustive in all aspects, yet it will constitute a tipping point.

Avoiding Collateral Censorship

One pressing question is whether the DSA could potentially be a means of collateral censorship (Balkin 2014). Indeed, this type of regulation can be considered as a way to impose content-related rules, although, from a constitutional law perspective, speech-restricting laws should be kept to a strict minimum. Some argue that the exception to Art. 14 e-Com-Dir, that is, Art. 6 DSA, could be a threat to freedom of expression because it incentivizes intermediaries to act against illegal content and, potentially, their own content rules (Kuczerawy 2021). This viewpoint builds on the over-removal phenomena, when platforms enforce more rules than necessary to avoid liability and the additional costs of nuanced content moderation practices (Keller 2019). While the risk of extensive enforcement of the DSA is a point to be taken seriously, the current draft clearly builds on balancing intermediary liability and fundamental rights under Art. 12 (2) DSA. This provision does not introduce a horizontal effect of freedom of expression between platforms and users, but it makes it mandatory to enforce only clear and unambiguous content rules (Kuczerawy 2021). A stricter liability regime would most certainly lead to the unwanted effect of collateral censorship (Buiten et al. 2020, p. 161). Ultimately, intermediary liability for illegal content is a constant dilemma (Helberger et al. 2018, p. 2; Heldt 2020).

CONCLUSION

More duties, more oversight, more transparency, and a systemic approach—the current proposal of the DSA provides answers on several levels. It addresses a wide range of issues and builds in safeguards at the individual and collective levels. Will it become the rulebook of reference for the digital sphere? It remains to be seen to what extent the final version of the DSA will contain crucial provisions or if the upcoming negotiations will delude them. One thing, however, is clear: the times of self-regulation are over—at least in the EU.

The DSA and other upcoming EU regulations could herald a new period for digital platforms and indirectly for users worldwide due to another perpetuation of the "Brussels effect" (Bradford 2012, 2020). According to Bradford, the EU developed a strong regulatory power at a global scale through its legal institutions and standards. Indeed, the EU aims for high standards in the Digital Single Market and could potentially develop what lawmakers consider a gold standard for platform regulation—beyond the EU's borders. This, however, presents them with another challenge: are they regulating for the EU or for the world (Heldt and Hennemann 2021)? In any case, one needs to also be aware of the developments across the Atlantic. On May 14th 2021, US-President Joe Biden revoked an Executive Order of former President Trump that targeted Sec. 230 CDA (Lyons 2021). Nonetheless, experts still expect the Biden administration to amend the current liability regime providing intermediaries with large immunity (Edelman 2021).

Meanwhile, the EU's main responsibility is to protect the European Union's values and rights (Art. 2 and 3 TEU), and, regarding "exporting" the DSA, there are two possible approaches. Either law-makers interpret this as an opportunity to develop the EU's power as a regulator beyond the EU's borders or, instead of generic rules which could be potentially adopted outside the EU, the DSA would be tailored-made for the EU and rely on rule-of-law guarantees provided by the Treaties. Hence, European lawmakers now might have to carefully gauge while keeping in mind that regulation like the DSA can be replicated by other countries.

REFERENCES

Balkin, J. M. (2014). Old School/New School Speech Regulation. *Harvard Law Review, 127*(8), 2296–2342.

Bradford, A. (2012). The Brussels Effect. *Northwestern University Law Review, 107*(1), 1–67. https://doi.org/10.1093/oso/9780190088583.003.0003

Bradford, A. (2020). *The Brussels Effect: How the European Union Rules the World*. Oxford University Press.

Buiten, M. C., de Streel, A., & Peitz, M. (2020). Rethinking liability Rules for Online Hosting Platforms. *International Journal of Law and Information Technology, 28*(2), 139–166. https://doi.org/10.1093/ijlit/eaaa012

Citron, D. K. (2014). *Hate Crimes in Cyberspace*. http://www.dawsonera.com/depp/reader/protected/external/AbstractView/S9780674735613

Citron, D. K. (2017). Extremist Speech, Compelled Conformity, and Censorship Creep. *Notre Dame Law Review, 93*(3), 1035.

Cornils, M. (2020). *Designing Platform Governance: A Normative Perspective on Needs, Strategies, and Tools to Regulate Intermediaries* (GOVERNING PLATFORMS, p. 88). AlgorithmWatch. https://algorithmwatch.org/wp-content/uploads/2020/05/Governing-Platforms-legal-study-Cornils-May-2020-AlgorithmWatch.pdf

De Streel, A., Defreyne, E., Jacquemin, A., Ledger, M., & Michel, A. (2020). *Online Platforms' Moderation of Illegal Content Online* (Policy Department for Economic, Scientific and Quality of Life Policies PE 652.718; p. 102). European Parliament. https://www.europarl.europa.eu/RegData/etudes/STUD/2020/652718/IPOL_STU(2020)652718_EN.pdf

Edelman, G. (2021, May 6). Everything You've Heard About Section 230 Is Wrong. *Wired*. https://www.wired.com/story/section-230-internet-sacred-law-false-idol/

Fanta, A., & Rudl, T. (2019, July 16). Leaked Document: EU Commission Mulls New Law to Regulate Online Platforms. *netzpolitik.org*. https://netzpolitik.org/2019/leaked-document-eu-commission-mulls-new-law-to-regulate-online-platforms/

Gillespie, T. (2014). The Relevance of Algorithms. In T. Gillespie, P. J. Boczkowski, & K. A. Foot (Eds.), *Media Technologies: Essays on Communication, Materiality, and Society* (pp. 167–194). The MIT Press.

Haltern, U. (2020). Ultra-vires-Kontrolle im Dienst europäischer Demokratie. *Neue Zeitschrift Für Verwaltungsrecht, 12*, 817–823.

Helberger, N., Pierson, J., & Poell, T. (2018). Governing Online Platforms: From Contested to Cooperative Responsibility. *The Information Society*, *34*(1), 1–14. https://doi.org/10.1080/01972243.2017.1391913

Heldt, A. (2019, April 16). Good Ends, Bad Means? The EU's Struggle to Protect Copyright and Freedom of Speech. *Council on Foreign Relations*. https://www.cfr.org/blog/good-ends-bad-means-eu-struggle-protect-copyright-and-freedom-speech

Heldt, A. P. (2020). Intermediärsregulierung: Quo Vadis NetzDG & Co? *UFITA*, *84*(2), 529–542. https://doi.org/10.5771/2568-9185-2020-2-529

Heldt, A., & Hennemann, M. (2021, July 15). Die Goldenen Zwanziger Jahre der Technikregulierung [The Golden Twenties of Tech Regulation]. *Tagesspiegel Background Digitalisierung & KI*. https://background.tagesspiegel.de/digitalisierung/die-goldenen-zwanziger-jahre-der-technikregulierung

Kaye, D. (2018). *Report of the Special Rapporteur on the Promotion and Protection of the Right to Freedom of Opinion and Expression* (Human Rights Council A/HRC/38/35). United Nations General Assembly. https://www.ohchr.org/EN/Issues/FreedomOpinion/Pages/OpinionIndex.aspx

Keller, D. (2019). *Who Do You Sue? State and Platform Hybrid Power Over Online Speech* (National Security, Technology, and Law, pp. 1–40) [Aegis Series]. Hoover Institution. https://www.hoover.org/research/who-do-you-sue

Kuczerawy, A. (2021, January 12). The Good Samaritan That Wasn't: Voluntary Monitoring Under the (Draft) Digital Services Act. *Verfassungsblog*. https://verfassungsblog.de/good-samaritan-dsa/

Llansó, E. J. (2020, August 17). The DSA: An Opportunity to Build Human Rights Safeguards into Notice and Action. *Medium*. https://medium.com/global-network-initiative-collection/the-dsa-an-opportunity-to-build-human-rights-safeguards-into-notice-and-action-by-emma-llans%C3%B3-e0487397646f

Lomas, N. (2020, December 30). Understanding Europe's Big Push to Rewrite the Digital Rulebook. *TechCrunch*. https://social.techcrunch.com/2020/12/30/understanding-europes-big-push-to-rewrite-the-digital-rulebook/

Lyons, K. (2021, May 15). Biden Revokes Trump Executive Order That Targeted Section 230. *The Verge*. https://www.theverge.com/2021/5/15/22437627/biden-revokes-trump-executive-order-section-230-twitter-facebook-google

O'Callaghan, D., Greene, D., Conway, M., Carthy, J., & Cunningham, P. (2015). Down the (White) Rabbit Hole: The Extreme Right and Online Recommender Systems. *Social Science Computer Review*, *33*(4), 459–478. https://doi.org/10.1177/0894439314555329

On the European Democracy Action Plan, no. COM(2020) 790 final, EU Commission. (2020). https://ec.europa.eu/info/sites/info/files/edap_communication.pdf

Schulz, W. (2019). *Roles and Responsibilities of Information Intermediaries: Fighting Misinformation as a Test Case for a Human Rights-Respecting Governance of Social Media Platforms* (Aegis Series No. 1904; National Security, Technology, and Law, p. 28). Hoover Institution. https://www.lawfareblog.com/roles-and-responsibilities-information-intermediaries

Schulz, W., & Dreyer, S. (2020). *Governance von Information*-*Intermediaren*— *Herausforderungen und Lösungsansätze* [Bericht an das BAKOM].

Scott, M., Kayali, L., & Vinocur, N. (2020, December 10). Tech Giants to Face Large Fines Under Europe's New Content Rules. *POLITICO.* https://www.politico.eu/article/tech-giants-to-face-large-fines-under-europes-new-content-rules/

Suzor, N. (2019, July 23). Lawless: The Secret Rules That Govern Our Digital Lives. *Medium.* https://digitalsocialcontract.net/lawless-2910ee226bfa

Tucker, J. A., Theocharis, Y., Roberts, M. E., & Barberá, P. (2017). From Liberation to Turmoil: Social Media And Democracy. *Journal of Democracy*, *28*(4), 46–59. https://doi.org/10.1353/jod.2017.0064

Wagner, B., & Janssen, H. (2021, January 4). A First Impression of Regulatory Powers in the Digital Services Act. *Verfassungsblog.* https://verfassungsblog.de/regulatory-powers-dsa/

Zarouali, B., Dobber, T., De Pauw, G., & de Vreese, C. (2020). Using a Personality-Profiling Algorithm to Investigate Political Microtargeting: Assessing the Persuasion Effects of Personality-Tailored Ads on Social Media. *Communication Research*, 009365022096196. https://doi.org/10.1177/0093650220961965

CHAPTER 5

Holding the Line: Responsibility, Digital Citizenship and the Platforms

Lelia Green and Viet Tho Le

INTRODUCTION

"[T]he current context is now fundamentally different, involving the use of our platform to incite violent insurrection against a democratically elected government," pronounced Facebook founder Mark Zuckerberg in relation to Donald Trump's use of the platform. Following pro-Trump protesters' storming of the US legislature on 6 January 2021, tech giants Facebook and Twitter decided that Donald Trump should be locked

This paper is based on a conversation with the late Professor Tom O'Regan, and a seminar delivered by Lelia Green at Tom's invitation at the University of Queensland in June 2019. It is dedicated to Tom's memory with gratitude, acknowledging his role as Principal Supervisor of Lelia's Ph.D.

L. Green (✉) · V. T. Le
School of Arts and Humanities, Edith Cowan University, Perth, WA, Australia
e-mail: l.green@ecu.edu.au

V. T. Le
e-mail: tho.le@ecu.edu.au

© The Author(s) 2022
T. Flew and F. R. Martin (eds.), *Digital Platform Regulation*,
Palgrave Global Media Policy and Business,
https://doi.org/10.1007/978-3-030-95220-4_5

85

out of the social network accounts he operated. These actions intensified longstanding debates around the risks that digital platforms may pose to democratic processes (Colarossi 2021). Critics on both the left and the right of the political spectrum have sought to curtail the legal protections that shield internet platforms from being held liable for content posted by people using social media. More recently, also reflecting Facebook and Twitter's ban on Trump, the platforms are being questioned around their decisions to ban everyday citizens, too. The essential point here is what duty of care is owed by big corporations when platforms operate as integral elements of the public sphere, but cannot be held publicly accountable?

On one side of the argument, the ban applied to Trump operates as a kind of pre-censorship. It raises concerns regarding platforms' power to moderate online content, their capacity for censorship, and protections for free speech on the internet (Oxford Analytica 2021a, 2021b). The opposing viewpoint argues that big tech has the right to censor content under their terms of service, which operate effectively as a contract between the platform and the people who use it. From this perspective, the platforms should take responsibility for preventing the promulgation of hate speech and disinformation. The polarising of debate concerning Trump's social media accounts serves as an example that highlights regulatory gaps, and the blurring of policies, governing social media platforms. It also raises the issue of corporate social responsibility (CSR) in terms of digital platforms' handling of hateful and/or violent messages (Hern 2021).

The aforementioned controversies around power wielded by the tech giants align with related concerns about surveillance capitalism (Holloway 2019; Zuboff 2019), and human and civil rights (Wagner 2018; Katyal 2019). They reinforce long-standing and widespread concerns about personal information privacy and the datafication of society (Van Dijck 2014). Further, such issues are not restricted to the corporate sector and it has become increasingly apparent that the corporate sector is being co-opted into state surveillance practices. PRISM, for example, was a code name assigned to a project run by the United States National Security Agency (NSA) which collected information and data with the support of US internet companies. Leaks by Edward Snowden in 2013 provided ample evidence that Americans' online interactions, as well as digital data relating to non-US citizens, had been monitored, collected and shared by many US-based digital companies. These organisations provided NSA operatives with direct access to their servers, allowing the collection of personal information relating to billions of people around the globe (Landau 2013; Stoycheff 2016). More recently the Cambridge Analytica scandal, revealed by whistle-blower Christopher Wylie in 2018, demonstrated platforms' complicity in manipulating users' perspectives upon politics and democratic processes, potentially impacting actions and behaviours, ethics and political outcomes (Cadwalladr and Graham-Harrison 2018; Isaak and Hanna 2018). Such examples of information, or grey-zone, warfare (Hughes 2020) demonstrate platforms' significant potential to exacerbate social fragmentation and polarise voters along ideological lines (DiFranzo and Gloria-Garcia 2017).

This chapter explores whether and how civil society in Western democracies can require platforms to take greater responsibility for power they wield in informing democratic deliberation and debate. It asks: what changes would platform media need to make to 'take responsibility' in the digital landscape? Exploring existing regulation and legislation, the argument adopts a Corporate Social Responsibility (CSR) perspective to deliver a more robust engagement with human rights and digital citizenship, benefiting individual citizens, and the societies in which they live. It suggests that the platforms and companies supporting major social media sites be constructed as 'publishers', and required to take responsibility for harmful content they carry. Specifically, the discussion addresses the following questions:

(i) What changes would platform media need to make to 'take responsibility' in the digital landscape?

(ii) How might the future formation of corporate social responsibility support a more constructive, pro-social engagement with information, data and knowledge?

(iii) Can platforms be held responsible for upholding individuals' rights? and

(iv) Do emerging understandings of CSR support more responsible practices by tech platforms?

The 'Big Five' Digital Technology Companies

It comes as no surprise to learn that the world's 10 most valuable publicly listed corporations include the five largest Big Tech companies, all of which are US based: Google (Alphabet, since a 2015 restructure), Amazon, Facebook, Apple, and Microsoft (GAFAM) (Frost et al. 2019; Clement 2021). Together, these corporations comprise what is known as 'the Big Five' (Van Dijck 2020). Beyond market value, the Big Five have gained "rule-setting power" (Van Dijck 2020, p. 2), which is to say they operate as the gatekeepers for almost all the western world's online social traffic and economic activities. Their services influence the very texture of society and impact the processes of democracy. In other words, online platforms are at the core of significant social change and development. They affect—and effect—institutions, economic transactions, and social and cultural practices (Chadwick 2014; Van Dijck et al. 2018).

This is not to say that these five tech companies start from, or pursue, the same approaches to their business models. Facebook and Google are essentially advertising driven and package users' data to build market share. While they have very different corporate cultures and individual responses to regulatory intervention—most recently illustrated in their responses to the 2021 Australian News Media and Digital Platforms Mandatory Bargaining Code (Leaver 2021)—their income streams depend upon the commodification of their users' information. Arguably, Apple is predominantly a hardware tech corporation, with Microsoft the dominant player in the software area. Amazon, in contrast, is a digital distribution behemoth, but it's also willing to enter niche markets (such as the carriage of high-end food retailer Whole Foods) in order to gather more data about particularly wealthy groups of customers, as well as to trial new modes of delivery. Since all these conglomerates are data-driven, there is some business-model slippage across products and services. Amazon's Alexa is an information-gathering device with a business model more aligned with Facebook and Google, while Apple TV + is more about cementing a hardware market than it is about taking on Netflix. There is some less-than-friendly rivalry between the Big Five. As this chapter goes to press, for example, Apple is in contestation with Facebook about its use of users' information. Apple has highlighted the different approaches by explicitly seeking users' consent to share some of their information with third parties (Statt 2021).

The Big Five have all developed from a single 'big idea' into huge conglomerates of interconnected platforms, going on to become dominant market players. Historically they built a core product, established its popularity and quickly disseminated it, adding value with aligned services and expanding operations to other sectors, while moving to dominate the market by acquiring potential competitors (e.g. Newton and Patel 2020). Given the cross-border, multi-market scale at which these companies operate, using national policies and laws to effect governance over these companies proves challenging. Further, the power and profit of these platforms operating in and across free market economies allows them to profit from outdated laws and inexact rules that fail the fit-for-purpose test when it comes to regulating digital environments and activities. The significant role that digital platform-driven companies play in the "heart of societies" (Van Dijck et al. 2018, p. 2) forces governments to second guess legal interventions, continually anticipating the next innovation and activity. Conventional regulatory approaches and instruments struggle to safeguard public interests (Nooren et al. 2018).

The European Commission, for example, has attempted a range of regulatory options, including self-regulatory and co-regulatory models (Finck 2017). A recent regulatory attempt proposed by the European Commission is founded upon "principles-based self-regulatory/co-regulatory measures, including industry tools for ensuring application of legal requirements and appropriate monitoring mechanisms" (2016). More recently, their regulation has been backed by significant sanctions and a threat of exclusion from one of the world's largest consumer markets (~450 million: the world's third largest population, after China and India). The EU's General Data Protection Regulation has been law since 2016 (Hoofnagle et al. 2019), and in force since 2018. It is supported by EU Regulation 2019/1150 promoting fairness and transparency for business users of online intermediation services (Anagnostopoulou 2020). These two regulatory tools aim to increase citizens' control of personal data and to protect civic society from the negative impacts of exploitative and predatory activities by digital platforms and services.

The General Data Protection Regulation changed the European privacy landscape (Hoofnagle et al. 2019) but also propted regulatory ripples worldwide. Among other reasons for this, the EU has a growing track record of enforcing regulation with respect to Big Tech. According to Keane (2015), Google is the world's largest and most dynamic media conglomerate and its revenue amounted to US$181.69 billion in 2020 (Johnson 2021a), with an operating income of US$49 billion in that year (Johnson 2021b). The platform may seem too big to regulate, but Google was subject to almost US$10 billion worth of fines between 2018 and 2020 for anticompetitive practice in the EU (Whalen 2020). Those kinds of penalty are one way to make platforms take notice. They also offer lawyers and regulators an opportunity to highlight the importance of a CSR ethics in the ways that platforms conduct themselves.

In October 2018 Facebook was fined £500,000 by UK regulators for its shortcomings as revealed in the Cambridge Analytica scandal. In July 2019 the platform, which includes Instagram, WhatsApp and Oculus, also settled a US Federal Trade Commission suit regarding Cambridge Analytica and other privacy issues, agreeing to pay a record-breaking US$5 billion fine while also implementing enhanced privacy measures (FTC 2019). This fine was big enough to see Facebook's net income drop in 2019, even though revenue increased from US$56 billion to US$71 billion (Tankovska 2021).

In 2018 Google, Facebook, Microsoft, Twitter and others, had joined together to form the Data Transfer Project (DTP 2018a), an initiative of the Google Data Liberation Front (a team of Google engineers), with the supposed aim of creating "an open-source, service-to-service data porta-bility platform so that all individuals across the web could easily move their data between online service providers whenever they want" (DTP 2018b). The idea was that an individual's content posted on Facebook, for example, could be seamlessly moved to Google + .

While such an initiative might sound impressive, and would be welcomed by many users, it is yet to be delivered. Further, the Data Transfer Project would not address the privacy and data control issues highlighted by the Cambridge Analytica scandal. Users remain subject to unregulated advertising that is driven by the Online Behavioral Adver-tising model (OBA) which underpins digital platforms such as Google and Facebook, providing 'free' services funded without explicit, informed consent by the monetisation of users' data (Edelman 2020; Torbert 2021). Arguably, given its progress, the Data Transfer Project is little more than window dressing to make the platforms appear to be doing more with respect to CSR ideals.

The operation of digital advertising/surveillance capitalism (Holloway 2019) belies any apparent improvements in platforms' ethical standards. It does more than construct audiences as "a commodity produced and sold to advertisers to use", Smythe's (1981) famous aphorism. OBA allows platforms to construct an image of a specific user's profile, forming what is termed 'like-minded audiences' articulated around features of specific importance to advertisers, including the shadowy covert operations influencing the Trump election campaign and the Brexit Referendum, both in

2016. Effectively a psychographic profiling technique, such 'digital experience' services are central to the targeted information delivery approach revealed in the Cambridge Analytica scandal, and integral to hidden, unregulated advertising. Users cannot capture the advertisements they have seen, interrogate them, or examine impacts upon them: which are essentially subliminal (Wachter 2020). This model of advertising operates without clarity or accountability, raising issues around the "overpassing [of] ethical limits in terms of respect for the persuadee, equity of the persuasive appeal, and social responsibility for the common good" (Belanche 2019, p. 685).

The Western policy agenda now reflects global concern around digital platforms' role and impact relating to the digital economy, privacy and personal data exploitation, misinformation and harmful content, etc., (Flew et al. 2020). Australia's Digital Platforms Inquiry report (ACCC 2019) is just one example of this concern, and particularly interrogates the impact of digital platforms upon consumer access to quality news and journalism.

This section of the paper has indicated that regulation, backed by sizeable fines, can help make platform media 'take responsibility' in the digital landscape (question i), and that corporate social responsibility, including around the regulation of the OBA model, could support a more constructive, pro-social engagement with information, data and knowledge (question ii). The EU's General Data Protection Regulation actions against Google, and the FTC's actions against Facebook, both indicate ways in which platforms may be held responsible for upholding peoples' rights (question iii). Question iv, 'Do emerging understandings of CSR support more responsible practices by tech platforms', is addressed in the sections that follow.

CORPORATE ENTITIES, CAPITALISM AND DEMOCRATIC IDEALS

The Australian Competition and Consumer Commission defines digital platforms as "applications that serve multiple groups of users at once, providing value to each group based on the presence of other users" (ACCC 2019, p. 41). The rapid growth of digital platforms highlights issues pertaining to CSR with an emphasis on the intersection between businesses, digital citizenship, and ways in which such entities are shaped by mutual interaction and mediated engagement with technology (Adi et al. 2015; Gold and Klein 2019; Schultz and Seele 2020; Stancu et al. 2018). The tech giants' operations necessarily raise issues requiring a CSR response (Grigore et al. 2018). A new CSR model for the digital age, where big tech companies face sanctions if they fail to adhere to a robust Code of Conduct, or an appropriate Code of Ethics, would add value to the implied commitment to CSR in digital discourse permeating the digital economy.

CSR has been defined as "an evolving business practice that incorporates sustainable development into a company's business model. It has a positive impact on social, economic and environmental factors" (Schooley 2020). Carroll (1991, 2016) suggests conceptualising it as a pyramid model constructed from four (deemed) constituent elements of CSR: Economic responsibility, legal responsibility, ethical responsibility, and philanthropic responsibility. There is no agreed definition of CSR, however. It operates as an umbrella term, in many senses as a buzzword or catch phrase, and is sometimes substituted for, or treated as if it were also referring to, environmental, social, and governance (ESG) aspects of corporate activity. Arguably, there are corporations that might continue to suggest that their only legitimate role is to maximise shareholder value. If they wish to have a social mandate to operate in a post-industrial information society, however, corporations need to be seen to be minimally ethical and avoid flouting standards of acceptable business behaviour. Flagrant disregard of public expectations can exact a significant toll on a company's balance sheet.

Beck (2019) argues that, nowadays, boycotts are a significant means of social protest against companies. Such boycotts can be called for in response to environmental pollution, violations of standards for workers, mistreatment of animals, etc. As a result, low CSR standards or performance have the power to undermine both profitability and share price, wiping out years of productive work to maximise shareholders' equity. In the alternative, positive CSR is perceived as supporting sustainability.

Consumers are increasingly aware of their buying power, and the value of their goodwill. Over the years they have become ever more inclined to call for, and participate in, mass boycotts. The 2015 Cone Communications/Ebiquity Global CSR Study found that 91% of global consumers expect companies to operate responsibly, with 84% saying that they seek when possible to consume goods made by responsible companies (Cone Communications 2015). On the investment side, 25% of organisations claim they operate in accordance with best practice standards of environmental, social, and governance principles (Flood 2019). The proportion of companies making such claims is expected to increase by more than double, to between 50 and 65% of all publicly reporting companies, by 2024 (Flood 2019).

When the Cambridge Analytica scandal broke, implicating Facebook in anti-democratic activities, that corporation lost US$45 billion in value over five days (*Economist* 2018). Although this value was subsequently regained, and retained, despite the FTC fine (FTC 2019; Davies and Rushe 2019), the initial precipitous drop in share valuation is a cogent indication of the risks that corporations run when they lose public trust. As a result of this and other examples, such as Rio Tinto's Juukan Gorge debacle (Verrender 2020), people working in finance and investments within western contexts cannot ignore the growing zeitgeist that mandates incorporation of CSR criteria into an evolving value equation. This dynamic also reflects the fact that low CSR commitment is an increasing regulatory and legislative risk. The Australian News Media and Digital Platforms Mandatory Bargaining Code, arising from the ACCC's Digital Platforms Inquiry (2019), is just one recent example. In a world first, it forced tech giants to pay Australian news outlets for their proprietary content when it is accessed, read and shared on social media and by search engine users.

The theoretical foundations of CSR are deeply interconnected with the idea of stakeholder engagement and, according to Freeman and Dmytriyev, it is "part of [the] corporate responsibilities oriented toward all stakeholders" (2017, p. 14). Carroll (1991) argues that, "the concept of stakeholder personalizes social or societal responsibilities by delineating the specific groups or persons business should consider in its CSR orientation" (p. 43). Such obligations impact digital platforms, as they do all other commercial entities. Platforms need to engage end users as well as investors. Given that digital platforms aim to build sustainable businesses, thereby taking economic responsibility, they also need to meet the expectations of their stakeholders, with a particular focus on two core categories of end-user – platform users/audiences and advertisers. Rieder and Sire conceptualise this process as a requirement for businesses to get stakeholders "on board" (2014, p. 199). For digital platforms, this means that the connection between CSR and stakeholders is, if anything, of greater importance because digital platforms operate in the context of a service industry, rather than providing tangible goods. In the same way that CSR forms a nexus for delivering social goods along with economic profits, so CSR connects stakeholders, markets, regulators and digital platforms.

Arguably, CSR has different implications for different market segments and operating conditions. Within the digital environment, CSR may imply that the platforms and related organisations operate to develop and support a conscious sense of an engaged citizenship, within the context of which the platform and its users work with each other to support democracy, free speech and principles of transparency and accountability. Facebook Australia's decision to restrict news publishing and sharing on 18 February 2021, in response to what the company perceived as an attack on its business model by requiring it to pay for the Australian-originated news content that users post on its platforms, constructed Facebook as an overpowerful bully. While Facebook may have characterised the precipitating introduction of the Australian News Media and Digital Platforms Mandatory Bargaining Code as an act of aggression, that regulatory action had far less perceived impact on the lives of everyday Australians than did the Facebook 'news ban' response (Hutchens 2021). Further, given that both Facebook and Google were impacted in equivalent ways, and Google reluctantly complied with the new regulations whereas Facebook (initially) countered and fought them, Facebook highlighted its response as out of proportion to the threat posed to it by Australia's regulators in the context of its global market dominance.

Facebook appears to have lacked a sense of the implied social licence under which it services Australia's social media discourse. In protesting the regulators' actions, it was perceived as harming "community groups, charities, sport clubs, arts centres, unions and emergency services" (Hutchens 2021). Facebook has always been more than a news source because of the operations of OBA. It provides a service that is created in the image of, and harnessed to the production of, information that's relevant to the interests of every Australian Facebook user, including: friends, families, communities, sports, arts, hobbies and health. It is a community space where ideas are shared and discussed. As well as showcasing news content, Facebook is often mined by news organisations for leads and stories. Further, Facebook's pages are used to confirm and contextualise what readers and viewers may have seen or heard elsewhere.

Based on the suggested nexus between CSR and stakeholder theory, news organisations and Facebook users are both key stakeholders. If one of the two groups is absent, the demand from the other reduces. If Facebook's aim is to build a sustainable product, it needs to recognise its responsibilty to the wider Australian community as well as to other groups of key stakeholders. In the end, this is what Facebook did, imperceptibly impacting their profits by negotiating with Australian news producers and supporting the coexistence and growth/sustainability of Australia's media and journalism industry. Facebook's temporary attempt to contravene the social contract, the implied CSR licence under which it operates, has been constructed as something akin to 'an own goal' in Soccer. As Lewis (2021) noted a week after Facebook's policy reversal: "the social network's hostile attack on Australian users reinforces the need to tackle the monopoly power of tech giants". A stronger commitment to CSR on Facebook's part would have allowed it to sidestep much of the opprobrium that followed, and would have left the iron fist unseen and unused in its velvet glove. As it was, the organisation opened itself up to wry comments about Facebook's agreement "to re-friend Australia" (Lewis 2021), and undermined public confidence in Facebook's understanding and performance of CSR.

CSR, Platforms and Regulators

Digital platforms comprising, among others, Facebook, Twitter, Google, Amazon, etc., have played a vital role in realising critical public values (Helberger et al. 2018) and making them more accessible. The absence of effective legislation and regulation governing the platforms is becoming more evident over time, however. Policymakers and lawmakers struggle to respond, trying to level up power and accountability differentials. Flew and Wilding (2021, p. 48) call it "the turn to regulation in digital communication." Grigore et al. (2018) suggest "a move from firm-centric orientations to stakeholder-centric orientations, and benefits and risks associated with the use of digital technology" (p. 24). Finck (2017) and Helberger et al. (2018) propose a co-regulation model to address the challenges inherent in cross-border multinational hegemonic organisations. In some ways, such a model recognises the operation of regional regulators attempting to work with and rein in international companies. Much of the newly enacted laws and regulations, in Europe as well as Australia, adopt this approach, making compliance with local law the price of doing business in the local market. In essence, this aligns local stakeholders' notions of CSR as being interconnected with organisations' best interests, thus explicitly linking the regulation of digital platforms to their licence to operate in key markets.

This section has indicated how emerging understandings of CSR are supporting more responsible practices by tech platforms, including the Big Five.

Competing Conceptions of Acceptability and Accountability

CSR, as it operates within the context of western democracies, is expected to align with the fundamental tenets of digital citizenship. Generally, attempts to regulate digital platforms begin with market-friendly self-regulatory and co-regulatory models and move along an interventionist scale to arrive at top-down legislative intervention (Finck 2017). The failure of platforms' self-regulation (Flew and Gillett 2020) is evident in examples such as Cambridge Analytica, both because such self-regulation not only lacks transparency but also because it does not account for the interests of actors other than those that benefit the platform itself (Finck 2017). Self-regulation is comparatively easy to ignore when problems arise that conflict with platforms' self-interest. Facebook, for example, claims to moderate the content posted on its site to prevent violence, pornography, and privacy violations but the boundaries between what is acceptable and prohibited is not always clear. In Vietnam, for example, Facebook may find itself pressured by state actors to remove or obfuscate dissent, which officials might deem as "undermining national security, social order and national unity" (Banyan 2013). This pressure exists when the content suppressed does not violate Facebook's publicised community standards. China, similarly, requires platforms to block content deemed illegal or offensive, and punishes platforms and services that don't comply. As Braw (2021) argues "For firms under pressure from China, it makes little sense to remain loyal to a home country where the share of revenue is often quite small if doing so brings the risk of losing a much bigger market." Many such state-issued regulations contrast with western ideals of free speech, however, where citizens may argue that platform review of content prior to posting is censorship, and anti-democratic (Gillespie 2017, 2018).

Finck (2017), and Helberger et al. (2018), accordingly propose co-regulation as an appropriate paradigm for future approaches whereby "companies develop […] mechanisms to regulate their own users, which in turn must be approved by democratically legitimate state regulators or legislatures, who also monitor their effectiveness" (Marsden et al. 2020, p. 1). This paradigm is also compatible with a CSR orientation that considers the benefits and risks to stakeholders of using digital technology (Grigore et al. 2018). It encourages CSR by promoting a better understanding of the challenges and risks that digital technologies might raise for stakeholder groups, not only for platforms themselves.

Such discussions take place in a context where has been "little reflection on the responsibilities of digital platforms in the markets in which they operate" (ACCC 2019, p. 1). Meanwhile, there is no clear agreement as to what comprises digital CSR, as the following discussion notes. Further, there is little regulation in smaller markets that is backed up by robust legislation that would encourage the Big Five platforms to change their behaviour. Ideally, a future-facing conception of CSR would embody the principles of open society, civic responsibility, market autonomy and accountability under the rule of law, as well as supporting an enhanced vision for digital citizenship, benefitting individuals, communities and the societies in which they live.

But what happens when democratic ideals clash in irreconcilable conflict? Such a contestation is highlighted by the example of the Christchurch shootings on 15 March 2019, when a gunman opened fire in two mosques in that New Zealand city, ultimately killing 51 people and injuring scores more. The gunman filmed his entire crime, posting it live on Facebook. The footage, which was subsequently copied and widely shared on social media, found its way onto the pages of some of the world's biggest news sites in the form of images, GIFs and even videos (Macklin 2019). Soon after the implications of the (re)posting were realized as a de facto part of the gunman's motivation, social media and news sites removed the images. In total, Facebook deleted about 1.5 million videos within the first 24 h of the attacks, automatically blocking a further 1.2 million upload attempts and removed 300,000 additional copies after they were posted (Macklin 2019). The event became a warning to platforms regarding their appropriation for terrorism and violence, and demonstrates the dark side of social media as a facilitator of xenophobia (Crothers and O'Brien 2020).

Jacinda Ardern, New Zealand's Prime Minister, drew upon models of world's best practice relating to suicide coverage, extrapolating that the airing of some information might create support for copycat behaviour (Greensmith and Green 2015). She also embraced emerging guidance around the reporting of mass shooters: don't name the shooter, don't discuss their politics, focus on victims, support stricken communities, and make change where possible such as banning the weapons and the transmission of the images. Arden is in one corner of a debate around how platforms should perform in terms of CSR. Two months' later, in Paris, Ardern joined with French President Emmanuel Macron to call for an end to "the circulation of abhorrent material." Seventeen countries and some tech companies, include Facebook, Twitter, Google, Microsoft and Amazon, responded to the 'Christchurch calling' by signing a pledge to stand against online terrorism and extremism.

Australia was deeply implicated in the Christchurch shooting. This was not only because of the very close trans-Tasman connection, but also because the killer was Australian, and Australia had failed to identify him as a terrorist threat (Tarabay and Graham-McLay 2019). In response to the killer's use of Facebook to publicise his crimes, Australian Prime Minister Scott Morrison said, amongst other things, that his country would do more to regulate international digital media companies. He suggested that organisations cannot be relied upon to do the right thing but require legislation. "It should not just be a matter of just doing the right thing. It should be the law," he said (Kelly 2019).

Jacinda Ardern has been widely praised for the intent behind the 'Christchurch call' and her demand that all footage of the Christchurch Mosque shootings be removed from the internet. In this case, there is a general agreement that images promoting violent hate crimes are unacceptable. There is a widespread uneasiness, however, about legislation that draws a line between what constitutes acceptable and unacceptable digital content. For example, a 2007 attack by a US Apache helicopter killed 12 people in Baghdad, Iraq, including two Reuters staff. The video of that atrocity was posted by WikiLeaks in 2010, calling attention to US forces' behaviour in the face of perceived threats posed by unarmed civilians. It stimulated debate about Chelsea Manning's and Julian Assange's right to publicise footage of US killings, and associated moral issues. These included whether the west was justified in subsequently allowing the screening of Daesh footage of executions (Schmid 2015). While Julian Assange argued for the legitimacy of his actions under a right to 'free speech' (Alexander and Stewart 2010), other moral issues raised include whether the Apache helicopter footage might have mobilised US public support for the end of the Iraq war and helped lead to "exit strategies" (Hasian Jr. 2012, p. 190).

If the public sentiment is that Jacinda Ardern was right to call for removal of the Christchurch mosque terrorist shootings footage, might the same arguments undermine Chelsea Manning's and Julian Assange's right to publicise a much shorter video documenting the killing of 12 civilians from a helicopter gunship? The question, as Rusbridger (2019) poses it, is: "Was it in the public interest that the world should have eventually seen the raw footage of what happened?". It may be relatively easy to justify access to Daesh footage as helping persuade western audiences that the organization is murderous, inhumane, and barbaric, thereby supporting military intervention (NATO 2015). That end may be argued as justifying those means. But trying to justify which media is widely publicised and which is not on the basis of 'motivation' for posting content is not a sound foundation for effective, unambiguous, enforceable regulation.

In a final example, from 2014, *The Australian* newspaper controversially published a front-page image of a seven-year-old Australian boy holding the head of a slain Syrian soldier given to him by his father. This was a touch paper for discussion about homebred terrorism in Australia (Klausen 2015). These cases highlight different aspects of what may or may not be socially responsible, what is or is not a defensible way to deal with media access to coverage of life and death in violent scenarios.

The above three case studies show the complexity of mandating digital platforms' adoption of CSR in deciding what constitutes good corporate digital citizenship. Is nuance possible? Judgement calls demand extraordinarily complex decision making to (say) justify the screening of an Apache helicopter attack 'in the public interest', but suppression of the Christchurch shootings under the same rationale. Such nuance goes to the heart of emerging understandings of CSR in support of what constitutes responsible practices by digital platforms.

CSR AND DIGITAL PLATFORMS: COMPLEXITY OR CHILD'S PLAY?

Prior to digitisation, organisations may have had time for decision-making around what is and what is not publishable in the public interest. In contemporary contexts, such decisions need to be made instantaneously, and are generally delegated to algorithmic computation. But can algorithms identify pro-liberal democratic priorities?

Western publics have focused on regulators' intentions to require the digital platform ecosystem to use its technology—from artificial intelligence, facial recognition software, biometrics, big data, machine learning, targeted communications, social media commentary, etc.—to make decisions in the public interest. Global discussions around this end include the General Data Protection Regulation in the EU, US Democrat Senator Elizabeth Warren's suggested breaking up of the tech giants, President Biden's recent assault on big tech's "anti-competitive practices" (Paul 2021), and the German government's legal measures against social media platforms that fail to take down hate speech, fake news, and defamatory content within 24 hours of it being posted. These are all battles over public values and competing social, economic and cultural interests.

This chapter has considered a range of critical incidents to draw attention to the issues raised by CSR in relation to digital platforms. These platforms are not rogue operators but neither are they entirely aligned with what an informed public might see as the ideal of supporting liberal democracies. Regardless of their influence in cultural and communication contexts, big tech companies are corporations run for a profit and designed to extract the greatest possible value from the workings of the 'free market' in late-capital societies.

What changes would platform media need to make to 'take responsibility' in the digital landscape? Evidence for effective intervention strategies, mainly from the EU, urges that people support new forms of CSR to enable an enhanced vision for digital citizenship that benefits both individuals and the societies in which they live. The remaining challenge is to embody democratic society's ideals, civic responsibility, and accountability under stakeholders' co-regulation and the rule of law. That is a possible way to combine a free market economy with an end to the unbridled commodification of citizens' data.

Joining Facebook, or using Google, costs people the data they use and produce. Implicitly, users agree to be monitored, but might it be possible to change this situation? Gillespie argues that:

> these platforms not only host that content, they organize it, make it searchable, and in some cases even algorithmically select some subset of it to deliver as front-page offerings, news feeds, subscribed channels, or personalized recommendations. In a way, those choices are the central commodity platforms sell, meant to draw users in and keep them on the platform, in exchange for advertising and personal data. (2018, p. 210)

The rights of users should be taken on board in stakeholder approaches to platforms' performance of CSR. Current regulations often construct digital platforms as a single category, such as communication, media, and e-commerce, rather than capturing different digital platforms' heterogeneity (Nooren et al. 2018). Furthermore, some large platforms are conglomerates of interconnected platforms (Nooren et al. 2018) with diverse characteristics. When Facebook bought Instagram, for instance, it was not just buying Instagram; it was closing down a potential competitor. Smyrnaios (2018) shows how platforms use vertical integration to support an internet oligopoly: "well positioned throughout the [value] chain, either through mergers or acquisitions, stock purchases, or exclusive and privileged partnerships with companies that are upstream or downstream of their core business" (p. 91).

Hard to measure benefits, such as the quality and diversity of services and products, require consideration (Coyle 2019). As Furman et al. (2019) believe:

> A pro-competition approach will provide a swifter and more proportionate means of addressing the competition challenges posed by the tendency of many digital markets to tip towards one or two large players. The introduction of a principle-based framework, developed in collaboration with the relevant players, is likely to be better suited than ex post enforcement to dealing with new and evolving practices in fast-moving digital markets. The presence of a stable and predictable framework would also provide welcome certainty to platforms on the rules of the game for operating in these markets. (pp. 123–124)

Takedown of child sexual abuse images and some aspects of violent and terrorist-related activity might be easily agreed as core business in western democracies. There is less consensus, however, on what constitutes hate speech, misinformation and tolerable forms of political debate even where it may be offensive and polarising; upon what is newsworthy and in the public interest, and what is not; and who or what should determine the boundary between these (Gillespie 2017). Addressing the 'regulatory imbalance' between traditional media and digital platforms (Flew et al. 2020), as reflected in the ACCC's Digital Platforms Inquiry (ACCC 2019), may offer one form of resolution. But patterns of information circulation and public use of digital media might suggest regulating digital platforms like news media agencies. As the ACCC (2019) notes:

Digital platforms actively participate in the online news ecosystem, performing several of the same functions as news media businesses. This means that digital platforms are considerably more than mere distributors or pure intermediaries in the supply of news content in Australia. Despite this, virtually no media regulation applies to digital platforms in comparison with some other media businesses. (p. 166)

Western publics' 'right to know' requires a nuanced balance of competing interests. Applying patterns of regulation, legislation and enforcement, and treating digital platforms like news media agencies, will potentially require digital platforms to pay more attention to CSR. In the case of the image of the seven-year-old Australian boy holding the head of a dead Syrian, *The Australian* was required to account for its decision to publish. Given that they contravened regulations and social norms, the paper had to advance an argument as to why publication was in the public interest. In news media contexts, the professional and ethical codes defining best practice play a crucial role in supporting responsible journalism (Donovan and Boyd 2021). Holding digital platforms to the same account as publishers and news agencies may support their more robust engagement with CSR.

Regulation (including self- and co-regulation), legislation and enforcement are all required if the platforms are to change their practices. Such changes will help make platform media take responsibility in the digital landscape, supporting a more constructive, pro-social engagement with information, data and knowledge. Platforms can and should be held responsible for upholding individuals' rights. Emerging understandings of CSR in the digital realm support improved operating practices on the part of tech platforms; but also by national and international agencies, and by regulators.

References

ACCC (2019). *Digital Platforms Inquiry: Final Report* (1920702474). https://www.accc.gov.au/publications/digital-platforms-inquiry-final-report.

Adi, A., Crowther, D., & Grigore, G. (2015). *Corporate Social Responsibility in the Digital Age.* Emerald Group Publishing.

Alexander, D. & Stewart, P. (2010). Leaked US Video Shows Death of Reuters' Iraqi Staffers, Reuters': US News, 6 April, https://www.reuters.com/article/us-iraq-usa-journalists-idUSTRE6344FW20100406.

Anagnostopoulou, D. (2020). The EU Digital Single Market and the Platform Economy. In *Economic Growth in the European Union* (pp. 43–57). Springer.

Banyan. (2013, August 9). Vietnam and the internet: The audacity of repression. *The Economist.* https://www.economist.com/banyan/2013/08/09/the-aud acity-of-repression.

Beck, V. (2019). Consumer Boycotts as Instruments for Structural Change. *Journal of Applied Philosophy, 36*(4), 543–559.

Belanche, D. (2019). Ethical Limits to the Intrusiveness of Online Advertising Formats: A Critical Review of Better Ads Standards. *Journal of Marketing Communications, 25*(7), 685–701.

Braw, E. (2021). How China Took Western Tech Firms Hostage, *Foreign Policy: Voice,* 19 January, https://foreignpolicy.com/2021/01/19/china-hua wei-western-tech-hostages-national-firms/.

Cadwalladr, C., & Graham-Harrison, E. (2018, March 18). Revealed: 50 Million Facebook Profiles Harvested for Cambridge Analytica in Major Data Breach. *Guardian.* https://www.theguardian.com/news/2018/mar/17/cambridge-analytica-facebook-influence-us-election.

Carroll, A. B. (1991). The Pyramid of Corporate Social Responsibility: Toward the Moral Management of Organizational Stakeholders. *Business Horizons, 34*(4), 39–48.

Carroll, A. B. (2016). Carroll's Pyramid of CSR: Taking Another look. *International Journal of Corporate Social Responsibility, 1*(1), 1–8.

Chadwick, A. (2014). The Hybrid Media System: Politics and Power. *Public Administration, 92*(4), 1106–1114.

Clement, C. (2021). Google, Amazon, Facebook, Apple, Microsoft (GAFAM)—Satistics & Facts [Website], *Statistica,* https://www.statista.com/topics/4213/google-apple-facebook-amazon-and-microsoft-gafam/.

Colarossi J. (2021, January 8). Banning Trump from Social Media Makes Sense. But Beware the Downside. http://www.bu.edu/articles/2021/trump-ban ned-from-twitter-facebook/.

Cone Communications (2015). *2015 Cone Communications/Ebiquity Global CSR Study.* https://www.conecomm.com/research-blog/2015-cone-com munications-ebiquity-global-csr-study.

Coyle, D. (2019). Practical Competition Policy Implications of Digital Platforms. *Antitrust Law Journal, 82*(3), 835–860.

Crothers, C., & O'Brien, T. (2020). The Contexts of the Christchurch Terror Attacks: Social Science Perspectives. *Kōtuitui: New Zealand Journal of Social Sciences Online, 15*(2), 247–259.

Davies, B., & Rushe, D. (2019). Facebook to Pay US$5bn Fine as Regulator Settles Cambridge Analytica Complaint, *The Guardian,* 24 July. https://www.theguardian.com/technology/2019/jul/24/facebook-to-pay-5bn-fine-as-regulator-files-cambridge-analytica-complaint

DiFranzo, D., & Gloria-Garcia, K. (2017). Filter bubbles and Fake News. *XRDS: Crossroads, The ACM Magazine for Students*, 23(3), 32–35.

Donovan, J., & Boyd, D. (2021). Stop the Presses? Moving from Strategic Silence to Strategic Amplification in a Networked Media Ecosystem. *American Behavioral Scientist*, 65(2), 333–350.

DTP (2018a). Data Transfer Project Overview and Fundamentals, *Data Transfer Project*, https://datatransferproject.dev/dtp-overview.pdf.

DTP (2018b). About Us [website tab], *Data Transfer Project*, https://datatransferproject.dev/.

Edelman, G. (2020). Why Don't We Just Ban Targeted Advertising? From Protecting Privacy to Savig the Free Press, It May Be the Single Best Way to Fix the Internet, *Wired: Ideas*, https://www.wired.com/story/why-dont-we-just-ban-targeted-advertising/.

European Commission (2016). *Communication on Online Platforms and the Digital Single Market Opportunities and Challenges for Europe*. https://ec.europa.eu/digital-single-market/en/news/communication-online-platforms-and-digital-single-market-opportunities-and-challenges-europe.

Finck, M. (2017, June 20). *Digital Co-regulation: Designing a Supranational Legal Framework for the Platform Economy*. LSE Legal Studies Working Paper No. 15/2017, https://ssrn.com/abstract=2990043 or http://dx.doi.org/10.2139/ssrn.2990043.

Flew, T., & Gillett, R. (2020, July). *Platform Policy: Evaluating Different Responses to the Challenges of Platform Power* [Paper Presentation]. International Association for Media and Communication Research (IAMCR). Tampere, Finland. https://iamcr.org/node/13261.

Flew, T., & Wilding, D. (2021). The Turn to Regulation in Digital Communication: The ACCC's Digital Platforms Inquiry and Australian Media Policy. *Media, Culture & Society*, 43(1), 48–65. https://doi.org/10.1177/0163443720926044.

Flew, T., Gillett, R., Martin, F., & Sunman, L. (2020). Return of the Regulatory State: A Stakeholder Analysis of Australia's Digital Platforms Inquiry and Online News Policy. *The Information Society*, 1–25. https://doi.org/10.1080/01972243.2020.1870597.

Flood, C. (2019, May 20). ESG Accelerates into the Investment Mainstream. *Financial Times*. https://www.ft.com/content/195232e7-07b7-36e3-a768-b8c63b6cc3fc.

Freeman, R. E., & Dmytriyev, S. (2017). Corporate Social Responsibility and Stakeholder Theory: Learning from Each Other. *Symphonya. Emerging Issues in Management*, 1, 7–15.

Frost, J., Gambacorta, L., Huang, Y., Shin, H. S., & Zbinden, P. (2019). BigTech and the Changing Structure of Financial Intermediation. *Economic Policy*, 34(100), 761–799.

FTC (2019). FTC Imposes $5 Billion Penalty and Sweeping New Privacy Restrictions on Facebook [Website], *Federal Trade Commission*, https:// www.ftc.gov/news-events/press-releases/2019/07/ftc-imposes-5-billion-pen alty-sweeping-new-privacy-restrictions.

Furman, J., Coyle, D., Fletcher, A., McAuley, D., & Marsden, P. (2019). *Unlocking Digital Competition—Report of the Digital Competition Expert Panel*. www.gov.uk/government/publications.

Gillespie, T. (2017). Governance of and by Platforms. In Burgess, J., Marwick, A., & Poell, T. (Eds.). *The SAGE Handbook of Social Media* (254–278). Sage.

Gillespie, T. (2018). Platforms Are Not Intermediaries. *Georgetown Law Technology Review*, *2*(2), 198–216.

Gold, M. K., & Klein, L. F. (2019). A DH that Matters. In *Debates in the Digital Humanities* (pp. ix–xiv). University of Minnesota Press.

Greensmith, G., & Green, L. (2015). Reporting Mass Random Shootings: The Copycat Effect? *Ethical Space: The International Journal of Communication Ethics*, *12*(3/4).

Grigore, G., Molesworth, M., & Farache, F. (2018). When Corporate Responsibility Meets Digital Technology: A Reflection on New Discourses. In Grigore, G., Stancu, A., & McQueen, D. (Eds). *Corporate Responsibility and Digital Communities* (pp. 11–28). Springer.

Hasian Jr, M. A. (2012). Watching the Domestication of the Wikileaks Helicopter Controversy. *Communication Quarterly*, *60*(2), 190–209.

Helberger, N., Pierson, J., & Poell, T. (2018). Governing Online Platforms: From Contested to Cooperative Responsibility. *The Information Society*, *34*(1), 1–14.

Hern, A. (2021, January 12). Opinion Divided Over Trump's Ban from Social Media. *Guardian*. https://www.theguardian.com/us-news/2021/jan/11/opinion-divided-over-trump-being-banned-from-social-media.

Holloway, D. (2019). Surveillance Capitalism and Children's Data: The Internet of Toys and Things for Children. *Media International Australia*, *170*(1), 27–36.

Hoofnagle, C. J., van der Sloot, B., & Borgesius, F. Z. (2019). The European Union General Data Protection Regulation: What It Is and What It Means. *Information & Communications Technology Law*, *28*(1), 65–98.

Hughes, G. (2020). War in the Grey Zone: Historical Reflections and Contemporary Implications. *Survival*, *62*(3), 131–158.

Hutchens, G. (2021). The Facebook News Ban Revealed How Problematic It Is to Rely on Corporations to Provide Fundamental Public Services, *ABC News: Analysis*, 20 February https://www.abc.net.au/news/2021-02-21/when-facebook-banned-news-australia-we-saw-role-it-plays/13175698.

Isaak, J., & Hanna, M. J. (2018). User Data Privacy: Facebook, Cambridge Analytica, and Privacy Protection. *Computer*, *51*(8), 56–59.

Johnson, J. (2021a). Google: Global Annual Revenue from 2002 to 2021 [Website], *Statistica*, https://www.statista.com/statistics/266206/googles-annual-global-revenue/.

Johnson, J. (2021b). Annual Operating Income of Google from 2013 to 2021 [Website: NB error in table title, correct in commentary], *Statistica*, https://www.statista.com/statistics/513129/operating-income-google/.

Katyal, S. K. (2019). *Private Accountability in the Age of Artificial Intelligence.* U.C.L.A Law Review, *66*(1), 54–141.

Keane, J. (2015). Why Google is a Political Matter. *The Monthly, 112*, 2433.

Kelly, L. (2019). Australia Plans Tougher Socia Media Laws for Failing to Thwart Violent Content, *Reuters World News*, 30 March, https://www.reuters.com/article/uk-newzealand-shooting-australia-laws-idUKKCN1RB00L.

Klausen, J. (2015). Tweeting the Jihad: Social Media Networks of Western Foreign Fighters in Syria and Iraq. *Studies in Conflict & Terrorism, 38*(1), 1–22.

Landau, S. (2013). Making Sense from Snowden: What's Significant in the NSA Surveillance Revelations. *IEEE Security & Privacy, 11*(4), 54–63.

Leaver, T. (2021). Going Dark: How Google and Facebook Fought the Australian News Media and Digital Platforms Mandatory Bargaining Code. *M/C Journal, 24*(2).

Lewis, P. (2021). Facebook's Capitulation in Australia Is the Beginning of He Project to Regulate Big Tech—Not the End, *The Guardian*, 23 February, https://www.theguardian.com/media/2021/feb/23/facebooks-capitulation-in-australia-is-the-beginning-of-the-project-to-regulate-big-tech-not-the-end.

Macklin, G. (2019). The Christchurch Attacks: Livestream Terror in the Viral Video Age. *CTC Sentinel, 12*(6), 18–29.

Marsden, C., Meyer, T., & Brown, I. (2020). Platform Values and Democratic Elections: How Can the Law Regulate Digital Disinformation? *Computer Law & Security Review, 36*, 105373.

NATO (2015). *Daesh Communications Campaign and Its Influence.* Riga, Latvia: NATO Strategic Communications Centre of Excellence. https://stratcomcoe.org/cuploads/pfiles/daesh_public_use_19-08-2016.pdf.

Newton, C., & Patel, N (2020). 'Instagram Can Hurt Us': Mark Zuckerberg Emails Outline Plan to Neutralize Competitors, *The Verge*, 29 July, https://www.theverge.com/2020/7/29/21345723/facebook-instagram-documents-emails-mark-zuckerberg-kevin-systrom-hearing.

Nooren, P., van Gorp, N., van Eijk, N., & Fathaigh, R. Ó. (2018). Should We Regulate Digital Platforms? A new Framework for Evaluating Policy Options. *Policy & Internet, 10*(3), 264–301.

Oxford Analytica (2021a). Big Tech's Anti-Trump Activism to Have Limited Impact. *Emerald Expert Briefings*. https://doi.org/10.1108/OXAN-DB2 58747.

Oxford Analytica (2021b). Power of Tech Firms Exposes Regulatory Gaps. *Emerald Expert Briefings*. https://doi.org/10.1108/OXAN-ES258725.

Paul, K. (2021). Biden targets big tech in executive order aimed at anti-competitive practices, *The Guardian*, 10 July, https://www.theguardian.com/us-news/2021/jul/09/joe-biden-executive-order-anticompetitive-big-tech.

Rieder, B., & Sire, G. (2014). Conflicts of Interest and Incentives to Bias: A Microeconomic Critique of Google's Tangled Position on the Web. *New Media & Society, 16*(2), 195–211.

Rusbridger, A (2019, May 26). US Efforts to Jail Assange for Espionage Are a Grave Threat to a Free Media. *Guardian*. https://www.theguardian.com/commentisfree/2019/may/26/prosecuting-julian-assange-for-espionage-poses-danger-freedom-of-press.

Schmidt, A. P. (2015). Challenging the Narrative of the 'Islamic State'. *ICCT Research Paper*, The Hague: The International Centre for Conter-Terrorism, https://icct.nl/app/uploads/2015/06/ICCT-Schmid-Challenging-the-Narrative-of-the-Islamic-State-June2015.pdf.

Schooley, S. (2020, June 26). What Is Corporate Social Responsibility. *Business News Daily*. https://www.businessnewsdaily.com/4679-corporate-social-responsibility.html.

Schultz, M. D., & Seele, P. (2020). Conceptualizing data-deliberation: The starry sky beetle, environmental system risk, and Habermasian CSR in the Digital Age. *Business Ethics: A European Review, 29*(2), 303–313.

Smyrnaios, N. (2018). *Internet Oligopoly: The Corporate Takeover of Our Digital World*. Emerald Group Publishing.

Smythe, D. W. (1981). On the Audience Commodity and Its Work. *Dependency Road: Communications, Capitalism, Consciousness, and Canada* (pp. 22–51). Norwood, NJ: Ablex.

Stancu, A., Grigore, G., & McQueen, D. (2018). Corporate Responsibility and Digital Communities: An Introduction. In *Corporate Responsibility and Digital Communities* (1–7). Springer.

Statt, N. (2021). Apple's Next iOS 14 Beta Will Begin Forcing Developers to Ask for Permission to Track You, *The Verge*, 28 January, https://www.theverge.com/2021/1/28/22253366/apple-app-tracking-transparency-opt-in-requirement-beta-launch.

Stoycheff, E. (2016). Under Surveillance: Examining Facebook's Spiral of Silence Effects in the Wake of NSA Internet Monitoring. *Journalism & Mass Communication Quarterly, 93*(2), 296–311.

Tankovska, H. (2021). Facebook's Revenue and Net Income from 2007 to 2020 [Website: NB error in table title, correct in commentary], *Statistica*, https://www.statista.com/statistics/277229/facebooks-annual-revenue-and-net-income/.

Tarabay, J. & Graham-McLay, C. (2019). Could the Christchurch attacks have been presented? *The New York Times*, 19 June, https://www.nytimes.com/2019/06/18/world/australia/new-zealand-terrorism-christchurch.html.

Torbert, P. M. (2021). "Because It Is wrong": An Essay on the Immorality and Illegality of the Online Service Contracts of Google and Facebook. *Journal of Law, Technology, & the Internet, 12*(1), 2.

Van Dijck, J. (2014). Datafication, Dataism and Dataveillance: Big Data Between Scientific Paradigm and Ideology. *Surveillance & Society, 12*(2), 197–208.

Van Dijck, J. (2020). Governing Digital Societies: Private Platforms, Public Values. *Computer Law & Security Review, 36,* 105377.

Van Dijck, J., Poell, T., & De Waal, M. (2018). *The Platform Society: Public Values in a Connective World.* Oxford University Press.

Verrender, I. (2020). Consequences for Rio Tinto Over Juukan Gorge Catastrophe Are the New Norm, *ABC News Analysis,* 14 September, https://www.abc.net.au/news/2020-09-14/superannuation-forcing-change-rio-tinto-juukan-gorge/12659824.

Wachter, S. (2020). Affinity Profiling and Discrimination by Association in Online Behavioral Advertising. *Berkeley Technology Law Journal, 35,* 367.

Wagner, B. (2018). Free Expression? Dominant Information Intermediaries as Arbiters of Internet Speech. In Moore, M., & Tambini, D. (Eds.). *Digital Dominance: The Power of Google, Amazon, Facebook, and Apple.* Oxford University Press.

Whalen, J. (2020). Europe Fined Google Nearly $10 Billion for Antitrust Violations, but Little Has Changed, The Washington Post, 10 November, https://www.washingtonpost.com/technology/2020/11/10/eu-antitrust-probe-google/.

Zuboff, S. (2019). *The Age of Surveillance Capitalism: The Fight for a Human Future at the New Frontier of Power.* Profile books.